Double Sunrise

Qantas Empire Airways
Indian Ocean Wartime Services
1943–1946

GEOFF GOODALL

DEDICATION

To my daughters Elizabeth and Katharine of whom I am extremely proud and who gave me such joy over the years

Double Sunrise

Qantas Empire Airways Indian Ocean Wartime Services 1943–1946

Geoff Goodall

ISBN: 978-1-7641937-3-3

Originally published 1979
This edition published 2025 by Avonmore Books in collaboration with Jane Goodall

Avonmore Books
PO Box 217
Kent Town
South Australia 5071
Australia

Phone: (61 8) 8431 9780

avonmorebooks.com.au

A catalogue record for this book is available from the National Library of Australia

Cover design & layout by Diane Bricknell

© 2025 Avonmore Books.

No part of this book may be reproduced or transmitted in any form or by any means, electronic or mechanical, including photocopying or recording, or by any information storage and retrieval system, without permission in writing from the publisher.

Front cover aircraft profile artwork by Juanita Franzi. Top to bottom: Liberator G-AGKU, Lancastrian G-AGLY and Catalina G-AGFM Altair Star.

Back cover photo: A 1944 view of the Qantas Empire Airways flying boat base on the Swan River at Nedlands, Perth. (Neil Follett Collection)

Contents

Glossary & Abbreviations 4

Qantas - What's in a Name? 5

Maps 7

Explanatory Note to this Edition 9

Preface 10

Introduction 12

Chapter 1   Setting the Scene 13

Chapter 2   An Alternative Indian Ocean Route 24

Chapter 3   Catalinas: Double Sunrise Service 33

Chapter 4   Liberators: The Kangaroo Service 67

Chapter 5   Lancastrians: Express Air Service 84

Chapter 6   Peace Comes to the Kangaroo Service 94

Appendix 1 110

Appendix 2 118

Appendix 3 119

Appendix 4 120

Appendix 5 123

Appendix 6 125

Sources 127

Index of Personnel 129

Glossary & Abbreviations

ANA	Australian National Airways
AOC	Air Officer Commanding
AUW	All up weight
BOAC	British Overseas Airways Corporation
CO	Commanding Officer
CofA	Certificate of Airworthiness
DAT	Directorate of Air Transport
DCA	Department of Civil Aviation
DFC	Distinguished Flying Cross
DFM	Distinguished Flying Medal
DSO	Distinguished Service Order
HMAS	His Majesty's Australian Ship
HMS	His Majesty's Ship
HP	Horse Power
IAL	Imperial Airways Limited
MAUW	Maximum All-Up-Weight
MLD	*Marineluchtvaartdienst* / Netherlands Naval Air Service
MPH	Miles per hour
OTU	Operational Training Unit
POW	Prisoner of War
QEA	Qantas Empire Airways
RAAF	Royal Australian Air Force
RAF	Royal Air Force
RPM	Revolutions per minute
RNZAF	Royal New Zealand Air Force
TAS	True Air Speed
TEAL	Tasman Empire Airways Limited
US	United States
USAAF	United States Army Air Force
USAF	United States Air Force
VHF	Very High Frequency
VJ Day	Victory over Japan Day

Qantas - What's in a Name?

The name Queensland and Northern Territory Aerial Service Limited was registered at Winton, Queensland, on 16 November 1920. The new company did not fly its first passenger service until 2 November 1922, from Longreach to Cloncurry in outback Queensland. It was Australia's second operational airline, following Western Australian Airways Limited, which commenced services along the Western Australian northern coastline on 5 December 1921.

The growing Q.A.N.T.A.S. business moved to Archerfield aerodrome in Brisbane and was registered as Qantas Limited. Scheduled services were flown within Queensland as well as providing aircraft for the Australian Aerial Medical Service (a predecessor to the Royal Flying Doctor Service), air taxi services, flying schools, aircraft sales and maintenance. The fleet was mostly De Havilland biplanes, including seven DH.50s constructed by the company at Longreach under licence.

Qantas became an international airline during 1934 when it won the Government tendering process to operate the Australian end of the inaugural England-Australia air mail service. The

The beginnings of Qantas: the Queensland and Northern Territory Aerial Service booking office at Longreach, Queensland, in 1923. (Qantas)

service was to be a joint operation with Britain's Imperial Airways Limited (IAL). A new company Qantas Empire Airways Limited (QEA) was formed on 18 January 1934, jointly owned by Qantas Limited and IAL (which was to become British Overseas Airways Corporation, now British Airways). Commencing in December 1934 QEA flew Brisbane-Darwin-Singapore with De Havilland DH.86 four-engined biplanes. At Singapore IAL took over the Singapore-London leg using a variety of biplane and monoplane airliner types.

From 1938 the Empire Air Mail Scheme introduced a flying boat operation, operated by both QEA and IAL using Short Empire all metal seventeen passenger flying boats. These brought a new level of passenger comfort but still took eight days of flying mostly by day with seven overnight hotel stops for the passengers. Among the changes initiated for the new flying boat service, the Australian terminus changed from Brisbane to Sydney, where a flying boat base was built at Rose Bay in Sydney Harbour and QEA moved its company headquarters to Sydney. QEA flew the Empire flying boats on the Sydney-Darwin-Singapore route, connecting with IAL at Singapore. Meanwhile the Qantas Brisbane base continued to maintain a network of scheduled services throughout Queensland and to Darwin.

On 3 July 1947 the Australian government purchased the British and Australian shareholding of QEA, and the airline became wholly Australian government owned. Management remained mostly the same and the operating name was not changed but the airline was now designated as the Australian international flag carrier.

In August 1967, to reflect the lessening British Empire, the company name was changed to Qantas Airways Limited. In August 1992 Qantas Airways absorbed the government-owned domestic carrier Australian Airlines (formerly Trans-Australia Airlines) adding a wide network of services within Australia to its established international routes. The following year the government privatised Qantas, stipulating that at least 51% of the shareholding must be Australian. In the turbulent years that followed, the airline retained the name Qantas Airways Limited and continued to grow and maintain its reputation for safety and quality as a world-class operator.

Because the events in this story occurred when the name was Qantas Empire Airways Limited, this book uses the terms QEA or simply Qantas.

The Qantas Limited logo in 1930.

The Qantas Empire Airways 1944 Kangaroo Service logo.

Maps

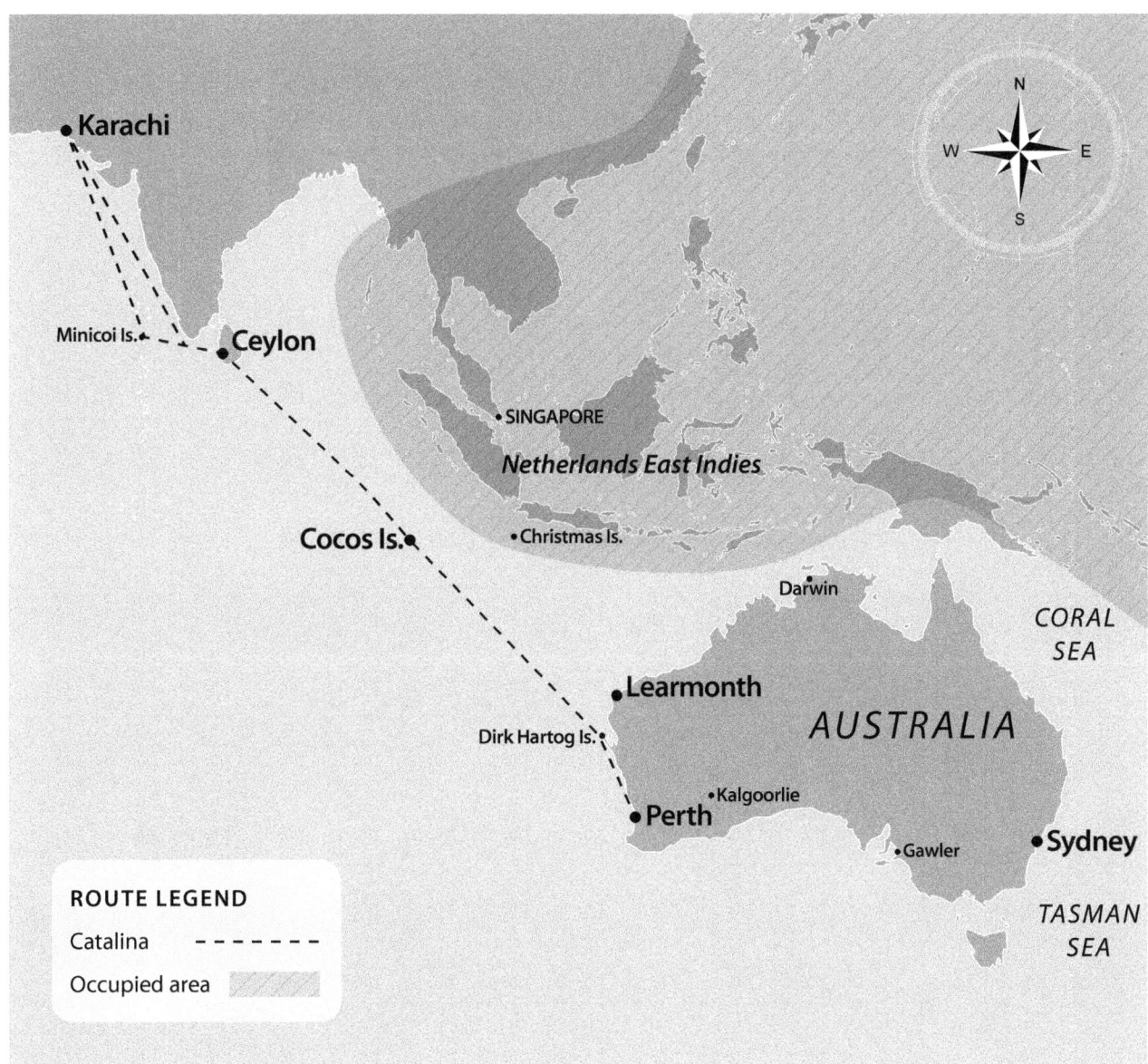

This map shows the wartime Indian Ocean routes flown by Qantas Catalinas. The sector Perth-Ceylon was 3,513 miles and was the longest non-stop airline route in the world. The Cocos Islands lay midway along the route and were overflown in darkness because patrolling Japanese aircraft were often active there.

The Perth-Ceylon Catalina service began in June 1943 and continued for over two years until July 1945. From November 1943 the service included an additional Ceylon-Karachi sector.

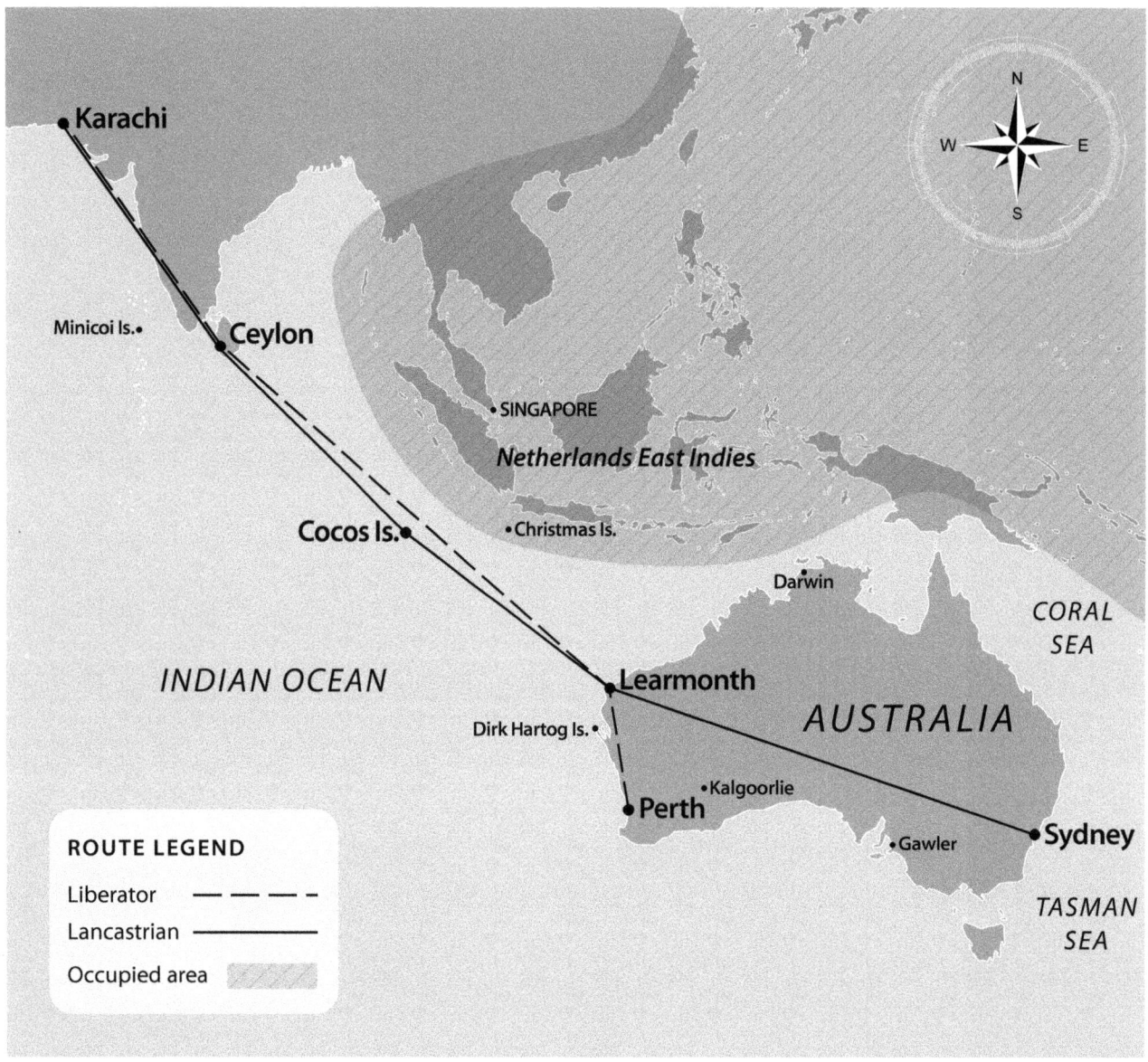

This map shows the wartime Indian Ocean routes flown by Qantas Liberators and Lancastrians. Liberator services began in June 1944, connecting Perth and Ceylon via Learmonth (Exmouth Gulf). An additional sector extended to Karachi.

Qantas began operating Lancastrians on the route from June 1945, and from November 1945 both Liberators and Lancastrians bypassed Perth altogether and flew Learmonth-Sydney direct. On occasions where adverse headwinds were experienced, a refuelling stop was available at Gawler in South Australia.

An RAF airfield had been built in the Cocos Islands in 1945. From January 1946 the Cocos Islands were used as a refuelling stop for Qantas Liberators and Lancastrians flying between Ceylon and Learmonth. Briefly a Cocos-Perth sector was flown until April 1946 when the Indian Ocean flights ended and the traditional Qantas route via Darwin and Singapore to Karachi was resumed.

Explanatory Note to this Edition

The original version of this book was written many decades ago by Barry Pattison and Geoff Goodall. Titled *Qantas Empire Airways (Western Operations Division) Indian Ocean Service 1943 – 1946*, it was published in 1979 by the Aviation Historical Society of Australia and printed by Qantas.

Research for the book in those days involved visits to physical archives and interviews with many of the aircrew and ground personnel involved with the service. In subsequent years Geoff Goodall continued his research, a task made easier with the internet era from the 1990s onwards. Indeed, Geoff had a broad interest in aviation history and maintained a website named *Geoff Goodall's Aviation History Site*.

In recent years Geoff prepared a manuscript for a revised and expanded version of the book, titled *Double Sunrise - Qantas Empire Airways Indian Ocean Wartime Services 1943-1946*. Geoff wrote a preface dated December 2021 but sadly the book had not been published at the time of his passing on 5 January 2024.

The book has now been published through the efforts of Geoff's widow, Jane Goodall. Geoff's 2021 preface is included in this edition, in which he acknowledges several people who helped with his research and manuscript. In addition, Jane wishes to acknowledge Tony Arbon, Ian McDonell and Nigel Daw who have assisted her with the publishing process. Thanks also to Ron Cuskelly for maintaining Geoff's website (goodall.com.au) which has also been archived by the National Library of Australia.

This edition includes Geoff's revised manuscript in entirety except for two appendices. These were very large files, comprising tabulated movement records for Qantas Catalinas and Liberators, and it was impractical to include them in a printed book. Instead, they have been uploaded to *Geoff Goodall's Aviation History Site*.

Preface

This is a revised and expanded edition of the original *Qantas Empire Airways (Western Operations Division) Indian Ocean Service 1943-1946* published in 1979 by the Aviation Historical Society of Australia. Barry Pattison, my co-author of that book had wanted to work together to produce this updated edition, but unfortunately his poor health has precluded another co-authorship.

During our original research we interviewed a number of retired Qantas personnel who had key roles in the secret wartime Indian Ocean services. These included Captain Bill Crowther, Captain Lewis Ambrose and Senior Engineer Norm Roberts, who welcomed us into their homes. The wealth of detail from their first-hand recollections, logbooks and records was the catalyst for publishing the first book.

This is a remarkable story of long-distance flying by the Australian airline Qantas Empire Airways Limited (QEA) during the Second World War, when the air route between Australia and Great Britain was cut by the Japanese occupation of the Netherland East Indies, Singapore, Malaya and Burma. To bypass enemy territory, a new route was proposed from Perth, Western Australia, to Ceylon but the long Indian Ocean crossing was beyond the endurance of military landplanes in the early days of the Pacific War. The result was the Australian and British governments tasking QEA to create a clandestine air service using British-supplied Catalina flying boats.

The flying time between Swan River, Perth, and Koggala Lake, Ceylon, averaged between 25 and 30 hours, flown in radio silence carrying diplomatic, military and war cabinet inter-government despatches and airmail. Because of the massive fuel load, a maximum of only three high-priority passengers could be carried. They endured an uncomfortable, cold and shatteringly noisy journey, during which they saw the sun rise twice. QEA produced a light-hearted paper certificate *The Secret Order of the Double Sunrise* which was signed by the aircraft captain and handed to each passenger at their destination. Little did the QEA aircrew realise that the magnitude of what they accomplished would be looked back on with the greatest respect today. Indeed, the "Double Sunrise" service is more widely known now than during the war years when it operated in secrecy.

Because many hours of each ocean crossing were flown through airspace where enemy aircraft might be encountered, the operation remained secret until the war situation improved. Most flights carried military personnel and military supplies. The civilian aircrew wore standard RAAF issue uniforms, with QEA rank and insignia. They were sworn in as members of the RAAF Reserve and were briefed that if captured by the enemy they were to give only their name, airline rank and RAAF Reserve number.

Other published accounts of the QEA wartime services have quite correctly acknowledged the extraordinary efforts of Captains Crowther, Ambrose and Tapp, however a fourth QEA officer must be included: Perth Senior Station Engineer Norman W Roberts. His determination to keep the aircraft flying and his personal resourcefulness was legendary. Indeed, after the publication of the original edition of this book, the Perth manager of Shell Oil at the time, Dudley Barker, wrote regarding Roberts:

> I cannot conceive that any man could have achieved more, in both the technical aspects and administration of the operation. That was despite injuries from a serious motor accident, which could have caused his termination from Qantas.

I have made extensive use of direct quotes. I

believe a quoted first-person account has far more credibility than an author's interpretation. All times quoted are local time. Locations and countries are referred to by their names at the time of the events described and similarly, weights and measures are given in pre-decimal figures.

This narrative has grown considerably thanks to input along the way from many of my aviation history colleagues. Thank you all. My special gratitude goes to three friends who made exceptional efforts to assist me with this book: Phil Vabre, Bob Livingstone and Paul Sheehan. Thank you for your willing help and expertise.

I hope this account brings attention to the magnificent efforts of all those engaged on the Qantas Indian Ocean services from 1943 to 1946.

Geoff Goodall
Melbourne
December 2021

Introduction

For two years from June 1943, QEA operated the world's then longest non-stop airline sector from Perth to Ceylon, a distance of 3,513 miles. Initially services were operated by Consolidated PBY-5 Catalina flying boats, whose average flight time was 27 hours. The longest crossing time was 31 hours and 45 minutes.

Qantas Captain William H Crowther, who was placed in charge of Qantas Western Operations Division in Perth to establish this secret wartime operation, wrote in a company report soon after the first few Catalina services:

> No other regular service anywhere in the world operates over anything like the 3,500 miles distance, the longest runs being San Francisco-Honolulu at 2,400 miles and Montreal-Scotland, a distance of 2,870 miles, but only flown direct when there is a following wind. Also occasional flights are done from Bermuda to the British Isles, a distance of approximately 3,100 miles. The service is, therefore, quite unique in the history of air transport. There never has been a service like it, and there probably never will be. It is secure already in its place of fame.

In June 1944 the Catalinas were supplemented by Consolidated Liberator transports flying a slightly shorter ocean crossing between Learmonth, Western Australia, and Ceylon. The Liberators brought a significant increase to loads, mail and passengers - plus a much more comfortable flight for passengers and crew. Later, Avro Lancastrians were also used on the route and, with the Liberators, flew the Indian Ocean route until April 1946 when Singapore reopened to airlines, allowing Qantas to revert to the far safer pre-war route via Darwin to Singapore. The name *Kangaroo Service* was coined when the Liberators commenced their Indian Ocean crossings and has since become synonymous with Qantas and its Australia-Britain services.

The story of the Indian Ocean service is one of triumph over adversity, manifested by the highest standards of airmanship and aviation endeavour. It took a special breed to crew the aircraft, for stamina was a requisite of the operation. Only one aircraft was lost operating the long ocean crossing; ironically it was a Lancastrian making one of the very final Ceylon-Learmonth Indian Ocean crossings. It was lost without trace on 24 March 1946, tragically claiming the lives of five Qantas aircrew and five passengers. Within weeks the Indian Ocean route was discontinued.

Chapter 1 – Setting the Scene

First Indian Ocean aerial crossing

When Australia joined the Empire Air Mail Scheme in 1934, the air route from Britain to Australia followed a geographic path between refuelling stops required by the short-range landplanes of the era. Imperial Airways flew the mail and passengers on a route across Europe, India, Burma and Malaya to Singapore. At Singapore, Qantas took over, the route generally being Singapore, Batavia, Surabaya, Rambang, Koepang, Darwin, Daly Waters, Cloncurry, Longreach, Charleville and Brisbane. When flying boats took over in 1938, the Australian terminus changed to Sydney, where the Department of Civil Aviation built the Rose Bay flying boat base on Sydney harbour. Qantas and Imperial Airways both operated the Short Empire flying boats handing over at Singapore as before. The Qantas flying boat route was Singapore, Klabat Bay, Batavia, Sourabaya, Bima, Koepang, Darwin, Groote Eylandt, Karumba, Townsville, Gladstone, Brisbane and Sydney.

During the late 1930s, Japan's growing militarisation and its invasion of Manchuria and China seemed remote from Australia. There was little thought given to the possibility that the Netherlands East Indies, Singapore, Malaya or Burma could be occupied by an enemy, thus cutting the air route to Britain. Although Qantas had seen first-hand Japanese encroachment in Portuguese Timor, in those heady days of empire at the government level the idea that "Fortress Singapore" could fall was unthinkable.

During 1935 a Perth entrepreneur, Charles LK Foote, commenced writing an increasingly strident series of letters to Perth newspapers predicting such a calamitous event and calling on the Australian government to finance an exploratory air route across the Indian Ocean. He also wrote to politicians and sent a detailed submission to the Minister for Defence. Frustrated by a lack of immediate response, Foote registered a company named Indian Ocean Air Services Limited in Perth, planning to use a Lockheed Vega fitted with floats to carry passengers from Onslow, Western Australia, to Mombasa via the Cocos Islands, Diego Garcia and the Seychelles. The minister responded that no approval would be given for a single-engine aircraft to fly such an oceanic route.

Behind the scenes, Australian government papers indicate that Foote's submissions were taken seriously by some departments. Input was sought from various defence committees and the United Kingdom government. By August 1937 the discussion had reached prime ministerial level when Australian long-distance record pilot and navigator Captain PG "Bill" Taylor threw his support behind an Indian Ocean aerial route survey. Taylor had flown fighters with the Royal Flying Corps in the First World War before returning to Australia to a civilian flying career. He had a strong interest in celestial navigation and was qualified in aerial navigation. During 1933 he was navigator and copilot for Sir Charles Kingsford Smith on Fokker F.VIIb *Southern Cross* on a series of flights to New Zealand. He also flew with Charles Ulm on Avro Ten VH-UXX *Faith in Australia* on record flights between Australia and Britain. In October 1934 Taylor and Kingsford Smith made the first eastbound crossing of the Pacific Ocean in Lockheed Altair VH-USB *Lady Southern Cross* from Brisbane to Oakland, California. These and his own solo long-distance flights made Captain PG Taylor a household name in 1930s Australia. He was highly placed in Sydney society and had an easy familiarity with politicians of the day.

While visiting England during 1938 Taylor

Captain PG Taylor (left) and Sir Charles Kingsford Smith, at Hawaii in October 1934 during their record-making eastbound crossing of the Pacific Ocean. (Powerhouse Museum)

personally lobbied the British Air Ministry and Short Brothers. Gaining press coverage back home, Taylor rode on that year's Australian 150th anniversary celebrations to write to the Government:

> My idea is absolutely based on it being carried out as an Australian Expedition, organised for the Australian Government in commemoration of the voyage of Captain Cook.

His formal proposal was submitted to Treasurer RG Casey on 7 October 1938, setting out all aspects of an Indian Ocean survey flight and offering his services. He nominated the Consolidated PBY as the most suitable aircraft and suggested that the PBY presently operating in New Guinea by the American Archbold Expedition might be chartered for the purpose.

The American Archbold Expedition in Netherlands New Guinea was completing its twelve-month flora and fauna survey. The full title was the American Museum of Natural History Expedition to Netherlands New Guinea, sponsored and led by wealthy New York zoologist Dr Richard Archbold. It was his third expedition to remote areas of Netherlands New Guinea and for this one he purchased a new Consolidated Catalina as a support aircraft. Newspapers reported that the Catalina was to be flown from Hollandia to Sydney for an overhaul by Qantas at Rose Bay before flying home to New York.

The expedition Catalina was a pre-military production Consolidated 28-3, equivalent to the US Navy model PBY-2 with minor differences. It was registered as NC777 on Archbold's request, as a replacement for his previous Consolidated 28-1 Catalina NC777. He also gave the newer aircraft the same name as the older one. This was *Guba*, a New Guinea Motu word meaning tropical storm.

Captain PG Taylor gained Australian Government support to approach Richard Archbold with a request to lease his Catalina for an Indian Ocean survey flight. The aircraft was to be operated by Archbold's crew, with Taylor in charge of navigation. The Australian government would meet all costs, including modifications at Rose Bay to install additional fuel tanks. The adventurer Archbold agreed, offering his aircraft and all facilities.

NC777 *Guba* with Doctor Archbold and his aircrew arrived at Rose Bay on 14 May 1939 where Qantas commenced work on an overhaul and requested modifications which were expected to take two weeks. PG "Bill" Taylor met Archbold and his flight crew to begin planning the flight. Taylor later wrote:

> The purpose I had originally conceived for this flight was to give a lead to direct air communication between Australia and Africa, and to explore island bases for a trans-Indian Ocean air service. The idea of an air service between Australia and Africa was, however, so far beyond the political and public view of the times that to obtain support for this flight I knew I would have to "sell" it on a much more visible objective. In my submissions

NC777 Guba on a November 1938 visit from New Guinea to the Rose Bay flying boat base, Sydney, for maintenance and to collect stores. The large hangar was under construction for the Qantas Short Empire flying boat service to England. (Ben Dannecker Collection)

to the Australian government, I therefore stressed the need for a reserve air route across the Indian Ocean which could be used to maintain air communication with the United Kingdom in the event of the Singapore route being cut by war and I proposed that we should make a survey of island bases over the region of that route.

Guba departed Sydney on 3 June 1939 for a nineteen-hour overnight ferry to Port Hedland on the northern coast of Western Australia. There the Catalina was laboriously refuelled on the harbour from cans of fuel lifted to the top of the wing from small boats. The group would commence the Indian Ocean survey flight the following morning. On board were Archbold and his five-man crew, together with Taylor and his Sydney newspaper journalist friend Jack Percival who would send reports to publicise the survey.

Taylor was impressed by Archbold's navigator, but on the first leg of the ocean crossing to the Cocos Islands, these two expert navigators, using the most advanced methods of marine and astro-navigation, were unable to locate the tiny island group. After some time searching, they were forced to divert north to Batavia in the Netherlands East Indies. After resting, a second attempt was made to reach Cocos and this time they were successful. They subsequently continued across the Indian Ocean to Africa, landing at Mombasa. There Taylor and Percival left the Americans, who continued on across Africa and the South Atlantic Ocean, reaching New York on 1 July 1940. The Indian Ocean stages flown on this historic flight were:

3-4 June 1939		
Sydney–Port Hedland	2,600 miles	19 hrs 35 mins
4-5 June		
Port Hedland–Batavia	2,025 miles	22 hrs 00 mins
7 June		
Batavia–Cocos Islands	700 miles	8 hrs 30 mins
13-14 June		
Cocos Islands–Diego Garcia	1,725 miles	14 hrs 17 mins
16 June		
Diego Garcia–Seychelles	1,022 miles	8 hrs 2 mins
21 June		
Seychelles– Mombasa	950 miles	7 hrs 40 mins

The crew of Guba at Sydney in May 1939, ready to depart on PG Taylor's Indian Ocean survey flight to Africa. Taylor is third from left, wearing a hat. (Bob Livingstone Collection)

Taylor submitted his detailed report to the Australian Prime Minister RG Menzies in Canberra on 2 September 1939, the day before war was declared. Taylor's report included surveys of operating areas for flying boats and potential airfield sites for land-based aircraft. It was immediately used by the British war cabinet for the selection of Royal Air Force and Royal Navy bases in the Indian Ocean for operations initially against German raiders - and after Pearl Harbor, against Japanese forces.

Qantas Spreads its Wings

Qantas became an international airline during 1934 when, amidst strong competition from other Australian airlines, it successfully tendered for the Brisbane-Singapore sector of the proposed England-Australia air mail service. Singapore to London was to be flown by Imperial Airways. Under the agreement, a new company was named Qantas Empire Airways Limited and was jointly financed by the two airlines. Both QEA and IAL ordered the new DH.86 Express ten-passenger biplane airliner, which promised increased speed and the safety of four engines.

The inaugural Qantas service to Singapore was scheduled to depart Brisbane in December 1934 but got off to an inauspicious start when the Australian Civil Aviation Board suspended the Certificates of Airworthiness of the Qantas DH.86s so far delivered. Inspections of their tailplane construction raised justified concern over structural integrity. Qantas was forced into a humiliating compromise of covering only Brisbane-Darwin with DH.50 and DH.61 biplanes, connecting with Imperial Airways in Darwin. The Australian DH.86 grounding caused outrage in Britain as it reflected badly on the British aircraft industry. The Civil Aviation Board lifted the type grounding in February 1935 after mandatory tailplane modifications were carried out on the Australian DH.86s. The affair had thrown a distressing cloud over the launch of the England-

Australia air mail service, which otherwise was a major aviation achievement.

The Qantas sector was extended to Singapore from 26 February 1935 when Captains WH Crowther and GU Allan departed Darwin in DH.86 VH-USC *RMA Canberra*, reaching Kallang aerodrome, Singapore, two days later. The England-Australia route now settled down to a reliable air mail service in both directions, only interrupted by occasional aircraft accidents. However, the limitations of the small DH.86 biplanes were recognised at government levels and moves were already under way to introduce large flying boats to replace the landplanes.

The type selected was the Short S.23 Empire designed by Short Brothers at Rochester, Kent. Imperial Airways took the unprecedented step of ordering 28 Empires off the drawing board and Qantas management placed orders for the same type. The Empires ("C Class" with Imperial Airways) proved a great success. Able to carry one and a half tons of mail and seventeen passengers, the Empires brought new levels of comfort to passengers on the England-Australia route. Advertising at the time showed well-dressed travellers standing watching passing scenery from the "promenade deck" – in fact a side aisle to the centre cabin with eye level windows and a handrail. The revised air mail service commenced with Empire flying boats on the England-Australia route from July 1938. The final QEA DH.86 service arrived at Singapore on 18 July 1938, when Qantas issued a press release stating that its DH.86s had flown 2,600,000 miles over the previous three and a half years, without any injury to passengers or crew.

With the introduction of the flying boats, the Australian terminus moved from Brisbane to Sydney. A new flying boat base was established at Rose Bay in Sydney Harbour. Passenger and maintenance facilities at Rose Bay were still being consolidated when the Second World War broke out in September 1939 and Prime Minister RG Menzies announced that Australia, as a loyal member of the British Empire, was therefore also at war. The Empire Air Mail Scheme was immediately suspended and flying boat services between England and Australia reduced as the Empire boats and crews were taken over by the RAF. Apart from this, hostilities in Europe had little marked effect on Qantas. However Imperial Airways, reorganised in 1939 as British Overseas Airways Corporation (BOAC), lost many European routes and from 1940 was forced to change the routing to Australia to a new "Horseshoe Route" from Durban, South Africa, via India, Singapore and Darwin to

Qantas Empire Airways DH.86 VH-USE as operated on the Brisbane to Singapore route in the 1930s. (Civil Aviation Historical Society)

Short S.23 Empire flying boats replaced the landplanes from 1938 on a revised Empire Air Mail Service. VH-ABF Cooee is pictured at the Rose Bay flying boat base, Sydney, in 1939. (Qantas)

Sydney. To assist BOAC, which was struggling to maintain services because of wartime aircraft and crew losses, Qantas extended its Sydney-Singapore sector to Karachi from October 1941.

Long Range Experience Gained

An Australian War Cabinet decision of 5 June 1940 approved the initial purchase of seven PBY-5 Catalina flying boats plus spare engines and parts for the Royal Australian Air Force. The order, quickly increased to eighteen aircraft, was placed through Brown & Dureau Limited, Melbourne, the Australian agents for both Consolidated Aircraft Corporation and Pratt & Whitney engines. Original plans to ship the Catalinas as sea freight, like the Lockheed Hudsons then being delivered to the RAAF, were scotched by shipping delays and the Catalinas' enormous 104-foot wingspan.

Because the United States was not yet at war, there were diplomatic considerations regarding the air delivery of military aircraft to an Allied country's air force. The US Department of State dictated that the Catalinas could be ferried to the last refuelling point in US territory in Hawaii by American civilian aircrew supplied by the manufacturer, where they could be taken over by Australian civilian crews for the rest of the delivery flight to Australia. There was no opposition to a civil delivery from the Department of Defence because the RAAF lacked capacity to undertake the task itself, having suffered from government peacetime complacency and budget cuts.

The Department of Civil Aviation (DCA) cooperated in the deception that these were civilian aircraft by allocating Australian civil registrations VH-AFA to VH-AFS for the ferry flights. These were a temporary allocation only and no Certificates of Registration were issued. The registrations were not applied to the aircraft, which were painted in camouflage, but each delivery flight used the civil registration in the official paperwork, and the fights were numbered "Trip A" to "Trip S" using the last letter of the registration. The Catalinas took up serials A24-1 to A24-18 upon arrival in Australia.

In early December 1940 Qantas Senior Engineer DH Wright left Sydney for San Diego to visit the Consolidated Aircraft Corporation factory. There he was briefed on all engineering aspects of the PBY-5 airframe and power plants, allowing him to prepare operating notes for Qantas aircrew. Lester Brain, PG Taylor and Qantas Senior Radio Officer AS Patterson followed on 28 December. They travelled by the Tasman Empire Airways flying boat to Auckland, then by the Pan American Airways Sikorsky Clipper service to Los Angeles. This gave them the opportunity to personally inspect the bases across the Pacific Ocean at which they would be stopping on the Catalina deliveries. Pan American extended every facility to the party, and they reached Los Angeles on New Years Day 1941.

In September 1940 the DCA approached Qantas requesting that the airline supply experienced aircrew with navigator's licences from the Sydney-Singapore Empire flying boat service. Qantas management, with thoughts of future expansion across the Pacific, willingly cooperated. The QEA Operations Manager, Captain Lester J Brain, was placed in charge of the operation. He formed the Long Range Operations Division expressly for the Catalina deliveries and selected captains, first officers, navigation officers and flight engineers. Until then, Qantas had not flown sectors longer than 800 miles, so Brain engaged Captain PG Taylor on a contract basis to oversee the navigation aspects of the Catalina deliveries.

Captain Lester Brain's company report takes up the story:

> On arrival at San Diego in early January 1941, I was informed by Mr FB Clapp, Australian Trade Commissioner in the United States, that the first Australian Catalina would not be ready until the first week of February. At the same time, I was informed that the first four British Catalinas were waiting on the ramp to be flown away, but the British delivery crew organisation could not cope at that time. Since the Australian and British machines were identical in practically every detail, even to camouflage and markings, it was suggested through Mr Clapp to the British authorities that we take over one of the four so we could return to Australia on schedule, and our first aircraft could be taken by the British when ready. Long negotiations followed and eventually the British authorities enquired whether I thought Australian crews available at San Diego could take two RAF aircraft immediately to Australia, one to remain the property of the Australian government and the other to be flown on to Singapore. Since Captain GU "Scotty" Allan lately with QEA and now with the RAAF and Mr FB Chapman, former navigator on the trans-Tasman service were in San Diego in addition to several RAAF radio men and engineers, I replied in the affirmative.
>
> The two aircraft were readied for departure on 24 January 1941. Owing to weather conditions and other matters, departure was delayed until the following afternoon. At 0830 on 25 January, the British authorities changed their minds and decided the second aircraft would not go to Singapore. Rapid rearrangement of crews was made.
>
> We adhered to the original plan, with Captain Taylor, Wright, Patterson and myself, with GU Allan and two RAAF sergeants as supernumeraries, accompanied by three Consolidated Aircraft Corporation aircrew. Captain O Denny was to join us at Honolulu. We took off at San Diego at 1540 local time on 25 January. In addition to the ten personnel there was baggage, emergency equipment and miscellaneous gear supplied with the aircraft from the factory. The take-off time was 47 seconds, and we took water over the cockpit for some seconds early in the run. Alighting at Honolulu 22 hours and 5 minutes later, we had sufficient fuel for a further 4 or 5 hours flying.[1]

There is considerable conjecture between aviation historians over the identities of the early Qantas deliveries. All indications are that the deliveries were as follows:

- Trip A VH-AFA: AH534 departed San Diego 25.1.41; arrived Sydney 2.2.41; to RAAF as A24-1.
- Trip B VH-AFB: departed San Diego 12.2.41; arrived Sydney 27.2.41; to RAAF as A24-2.
- Trip C VH-AFC: AH540 arrived Sydney 12.3.41; delivered to RAF Singapore 20-23.3.4.
- Trip D VH-AFD: departed San Diego 5.4.41; arrived Sydney 12.4.41; to RAAF as A24-3.
- Trip E to Trip S: VH-AFE to VH-AFS: A24-4 to A24-18 in sequence; final delivery October 1941.

Qantas crews collected the aircraft at Honolulu and flew them to the Rose Bay flying boat base in Sydney. There Qantas engineering staff carried out an inspection of each aircraft before RAAF crews delivered them to RAAF Rathmines on Lake Macquarie near Newcastle, New South Wales. The planned stages were:

Honolulu-Canton Island	1,921 miles
Canton Island-Noumea	1,900 miles
Noumea-Sydney	1,236 miles

---

[1] By the end of WWII, the RAAF had received a total of 168 Catalinas of various models.

PBY-5 AH534, diverted from a British order, at Honolulu in January 1941 after a 22-hour flight from the Consolidated factory in California under the command of QEA Operations Manager Captain LJ Brain, who continued the ferry flight to Sydney. (Lester Brain Collection)

The first delivery was only the third aircraft to cross the Pacific Ocean from the US to Australia. The deliveries were conducted without any losses and reflected highly on the abilities of the captains and crews involved, a number of whom were later to participate in the Indian Ocean operations. Of the 100,000 miles flown, over 40,000 miles were at night to allow for astronavigation. E Bennett-Bremner in his book *Front Line Airline* stated:

> The practically faultless running of the Catalina deliveries was a tribute to the organising ability of Qantas Operations Manager, Captain LJ Brain, who was responsible for the whole set of schedules. Almost without exception every senior skipper and first officer assisted in the deliveries and gained considerable experience in the operation of this, to the Company, a new type of aircraft. Wonderful work was done by the radio staff, and the Company's engineers also had the opportunity to add another aircraft endorsement to their ticket.

Only one significant schedule delay occurred, described by Qantas engineer Norm Roberts:

> There was only one delay, and I was the Flight Engineer involved on that occasion - the 18th delivery, A24-18, VH-AFS. This Catalina arrived in Pearl Harbor direct from San Diego with the starboard engine shut down. It had consumed most of the engine oil some four hours before landing at Pearl Harbor. We requested a new engine, and we asked for help, but the US Navy command declined because they were not at war. After several days I eventually met up with a Pratt and Whitney field representative and together we removed three suspect cylinders from the engine. In each of these we found wedge type rings had been fitted to square ring groove pistons and considerable wear to the lands had resulted. We checked the engine out completely, fitted three new cylinders and pistons, and the ensuing flight to Sydney was uneventful.
>
> This experience personally cautioned me from then onwards to monitor closely oil consumption rates during long flights. So it was that during the QEA Indian Ocean Catalina services, I called upon the RAAF at Boulder who did our engine overhauls to hold the wear tolerances within tight limits, to ensure oil usage remained minimal throughout the engine service life.

The final Catalina A24-18 arrived in Sydney on 22 October 1941 after staging via Canton Island and Suva, Fiji. It carried the first airmail between Fiji and Australia.

Captain HB "Bert" Hussey's pilot logbook gives a good indication of how QEA flight crews were rotated between the Sydney-Singapore service and the RAAF Catalina deliveries. Hussey was in command for eight Pacific ferry flights: A24-3, -4, -6, -8, -10, -12, -16 and -17.

During the Catalina ferry operation, some Qantas captains achieved flights over longer route stages, aiming to obtain the maximum range from the fuel available. These were carefully controlled, meticulously planned exercises, which became invaluable experience for the future Indian Ocean service. Captain Russell Tapp did best: he flew Canton to Sydney non-stop, overflying both Noumea and Fiji. His navigator on this 26-hour flight was PG Taylor and the aircraft was A24-15. The following is from Captain Tapp's flight report:

> It became the aim of each crew to overfly Noumea and do Canton Island to Sydney, 3,137 miles, non-stop. The normal range of the Catalina was approximately 2,300 miles, so that with a slight average tail wind component, and operating so that no power was wasted at any time of the flight, the distance of 3,137 miles could be flown. It was essential of course that suitable winds were found to make the flight and naturally they were not always existent.
>
> On October 4th 1941 three Catalinas left Pearl Harbor, one commanded by myself with Captain PG Taylor as navigator and John Connolly as first officer. The other two were under the command of Captains Crowther and Hussey. All arrived at Canton Island that night. Although we left Pearl Harbor together, we flew individually and lost sight of each other after a few hours' flying.
>
> Our plans were to take-off for Noumea and/or Sydney just before dusk the following evening. The following morning during a preliminary check of the weather ahead, it looked possible that conditions suitable for a non-stop flight might be found and we all discussed our chances of doing so. After an early lunch we turned in for a sleep. Crowther and I took off one behind the other and operated straight away to our long-range plan. This meant very careful engine operation and aircraft flying in order to avoid all wastage of power, and a careful study of all wind indications from cloud or water was necessary in order to take advantage of every bit of tail-wind component that we could find. Our positions were broadcast every hour, so it was a simple matter to know where the other aircraft was.
>
> The aircraft being absolutely new, we had, of course, checked the fuel gauge readings all along the first sector from Honolulu to Canton. The comparison of these readings after these tanks had been refilled told us exactly what fuel had been burnt. The type of gauge, however, is liable to slight fluctuations so that all along the flight we had maintained two "fuel remaining" sets of figures, one of the actual fuel gauge readings and one of what we calculated the amounts should be from the results of the previous flight. At times during the night there had been some differences between the two columns but on nearing Noumea, however, they were almost even. When we were just outside Noumea the gauges showed that we had just the absolute bare minimum left to continue to Sydney. There was ample, however, to continue to Brisbane. We decided not to land but to alter course slightly from Brisbane and to continue checking carefully, so that, if after a few more hours' flying the gauge readings were equal to our calculated fuel left and no unexpected headwind was met, we could alter course again for Sydney.
>
> We did not tell Crowther of our decision but observed silence until we heard his signal that he had landed at Noumea. By this time, of course, we had passed Noumea, and we then sent a signal giving our position and advising that we were continuing to Sydney. The flight onwards worked out according to plan and three hours after passing Noumea we altered course again for Sydney, arriving there at exactly 26 hours to the minute after leaving Canton Island. Our consumption during the last stages was down to 42 gallons an hour and, on dipping the tanks after landing, we found we had 84 gallons left, therefore having two hours' further fuel, exactly what we calculated it to be.

Japan's entry brought the war to Australia. Australian civil airliners were painted in camouflage. Short Empire VH-ADU at Rose Bay was the former BOAC G-AEUB Camilla. (John Hopton Collection)

Empire Air Mail Route Cut

Japan's entry into the war in December 1941 and rapid invasions of the Philippines, Malaya and the Netherlands East Indies dramatically changed Qantas operations. As Japanese forces advanced through Malaya towards Singapore and occupied large areas of the Netherlands East Indies, it was obvious that vital refuelling bases would fall into enemy hands, cutting the Empire air route.

Ten days before British command in Singapore surrendered to the Japanese, Captain WH Crowther operated the last Qantas departure. On 4 February 1942 at 0230 he made a hazardous night take-off from congested Singapore harbour, his Empire flying boat carrying 40 refugees, mostly British women and children. Crowther had seen the destruction and panic in the city caused by frequent Japanese air attacks. As he flew down the darkened coastline to Batavia, listening to the Japanese radio reports of his own departure, Crowther knew the fall of Singapore and the NEI was imminent. Air communication between Australia and Britain had now been effectively cut.

QEA had lost its international air route. Five Empire boats had already been chartered by the Australian government for issue to the RAAF, some being fitted with armament. Many QEA aircrew were already in the RAAF Reserve under the terms of the Empire Air Service Act and had been called up. A number of maintenance and support staff resigned to enlist in the military for the war effort. The airline's scheduled operations were limited to:

- Brisbane-Darwin landplane service, usually operated by Lockheed 10 VH-AEC *Inlander*.
- Sydney-Darwin government charter Empire flying boat weekly service from March 1942, increasing to twice weekly return from May 1942.
- Outback Queensland mail routes and flying doctor work with De Havilland Fox Moths and Dragons.

However, even these routes were cut back when Qantas lost these aircraft to impressment for the RAAF.

As Japanese forces invaded New Guinea and attacked northern Australia, the remaining Empire boats and other Qantas aircraft were frequently chartered by the government to fly rescue and supply missions into war zones. Late in 1942 some relief came with an agreement between the military Directorate of Air Transport and Australian airlines to operate military transport aircraft supplied by DAT. From November 1942 QEA was allocated two Lockheed Lodestars for a daily courier between Port Moresby, Townsville and Brisbane. Later three Douglas C-47s were operated on extended courier runs to Australian forces in Borneo. QEA maintenance bases at Archerfield and Rose Bay received much overhaul and repair work on Australian and US military aircraft.

Japanese advances made it impossible for BOAC's side of the "Horseshoe Route" to proceed further east than Calcutta. For the next eighteen months air communications between Australia and Britain relied solely on US military transports flying the much longer route from Sydney across the Pacific Ocean to California, then from Montreal across the Atlantic Ocean to Britain. The time taken to cover the extra distance was further increased by restricted airline bookings within the United States, which resulted in military priority passengers and diplomatic dispatches from Australia often travelling by rail from California to Canada. At Montreal they joined the RAF/BOAC North Atlantic return ferry service via a Liberator flight to Prestwick, Scotland, and then finally by train to London.

From the time Singapore fell to the Japanese, a senior Qantas pilot, Captain Crowther, was personally determined to plug the gap in the old route to England via India. In a written submission to Qantas founder and Managing Director, Hudson Fysh, he proposed a non-stop Consolidated Liberator service between the Australian west coast and Ceylon, then on to Karachi to join with the BOAC route now forced to terminate in India. The Qantas operations manager, Captain Lester Brain, supported Crowther's proposal and a series of detailed discussions were held with Fysh at the Sydney head office. Although the Liberator bomber stripped of armament and fitted with seating was the preferred type, it was realised that military priorities made it doubtful they would be released. It was agreed that Consolidated Catalina flying boats, although slower, had the long range required and would be suitable. On 27 April 1942, Fysh wrote to the Director-General of Civil Aviation in Australia, AB Corbett, setting out the proposal. Corbett, a strong-willed career public servant with no aviation background, replied on May 1 with a blunt cable:

The area is not regarded as safe for flying boats. The prospect of obtaining Catalinas is most remote. My reaction is that at present such a proposal would be little short of murder of pilots, and I would strongly oppose risking crews' lives.

Undeterred in his belief that Qantas could operate an Indian Ocean route, Hudson Fysh discussed the concept in his correspondence with senior management of partner airline BOAC, followed by a detailed proposal in October 1942. He arranged to visit Britain early the following year, receiving government approval for a priority air passage on the Consairway B-24 Liberator courier service from Sydney to California and onwards from Montreal to Prestwick, Scotland, by an RAF Ferry Command Liberator. Fysh was motivated by commercial as well as patriotic ideals: he had to keep Qantas functioning as an operational long-distance airline with a view to post-war regaining its position as Australia's international airline. There was no question in his mind as to Qantas' ability to operate across the Indian Ocean.

Early in 1943 BOAC submitted its own proposal to the British Air Ministry, suggesting a service be operated Perth-Ceylon to reinstate a direct air route between Australia and England. Their proposal was that Catalinas would be made available by Britain, to be flown by Qantas crews using the Cocos Islands as an enroute refuelling stop. When put to the Australian Government, this proposal was rejected on the grounds that Cocos was unsafe because of regular enemy reconnaissance flights. The idea for an Indian Ocean service lapsed for a brief period until the British Army high command and the Commander-in-Chief India and Burma, General Wavell, brought pressure on the British War Cabinet to re-establish a reliable air service directly between Britain and Australia.

Chapter 2
An Alternative Indian Ocean Route

Earlier Dutch Crossings

By March 1942 after battles in the defence of the Netherlands East Indies, the Netherlands Naval Air Service (*Marineluchtvaartdienst* or MLD) had only nine Catalinas remaining from a total of 36 received. Finally, they were forced to evacuate their last NEI base, Tjilatjap in Java, to avoid capture by the advancing Japanese.

Some Dutch Catalinas had already made their escape to north-western Australia during the hectic last days of February 1942, some carrying aircrew families. MLD Catalinas serials Y-49, Y-62, Y-69, Y-71 and one other (possibly Y-45) eventually reached the RAAF flying boat base at Rathmines near Newcastle, where a temporary NEI air group was established. However, four others (Y-59, Y-60, Y-67 and Y-70) were not so fortunate - they had reached the north coast of Western Australia at dusk on 2 March and alighted on the harbour at Broome. Because of demand on land facilities, their occupants were ordered to stay on board overnight. Next morning they were sunk at their moorings by a Japanese strafing raid from Timor, along with MLD Dornier Do 24s and other allied flying boats, with a terrible loss of life, both civilian and military. Broome aerodrome was also attacked, destroying Dutch, Australian and American bombers and transports.

On 1 March, six MLD officers and 29 airmen departed Tjilatjap in four battle-damaged Catalinas (Y-55, Y-56, Y-57 and Y-64) to find sanctuary in Ceylon. One aircraft, flown by Captain Mueller and Flight Lieutenant Wilhot, carrying 66 desperate refugees, could not reach Ceylon and landed unannounced at the Cocos Islands, seeking fuel. The hull was torn when they struck a submerged coral head while taxiing on the lagoon, flooding some interior compartments. The Catalina was pulled on to a beach by a large group of Malay workers, and

Netherlands Naval Air Service (MLD) PBY-5 Y-87. The red, white and blue flag replaced the Netherlands orange triangle nationality insignia for Dutch aircraft in the Far East to avoid confusion with the Japanese red disc. (Nationaal Archief Netherlands via AirHistory.net)

the crew patched the hull with a mixture of cement and washing soda before continuing to Ceylon.

During May 1942 the Commander-in-Chief, Netherlands Forces Eastern Command, Vice Admiral CEL Helfrich, ordered the five MLD Catalinas in Australia to proceed to Ceylon. They left Rathmines in June 1942 and proceeded to the US Navy flying boat base at Exmouth Gulf, Western Australia. Three then flew via the Cocos Islands to refuel, while the other two flew direct to China Bay, near Trincomalee, Ceylon, making the first non-stop crossing of the Indian Ocean from Australia to Ceylon. Their squadron commander, Commander van Prooyen said:

> At the time no Catalina had ever made this long sea crossing non-stop, but I very much wanted both men and aircraft at that particular time. On the day the flying boats reached Ceylon I anxiously scanned the horizon. At last I saw one coming, and it made a good landing. All five arrived.

The assembled Dutch Catalinas and crews in Ceylon were officially taken on RAF charge on 1 July 1942 and, as from 15 August, were assigned to the reformed No. 321 (Dutch) Squadron under RAF

No. 222 Group, Colombo. The Dutch squadron was established at RAF China Bay, Ceylon. The circumstances under which MLD crews made this long ocean crossing to Ceylon make this achievement all the more remarkable; the crews had fought to exhaustion in many battles, and there was no time to properly prepare for the long-distance flight.

In December 1942 a No. 321 Squadron Catalina made the first of three flights from Ceylon to Perth. Y-45, under the command of Lieutenant WJ Reijnierse with RAF Wing Commander L Fox, DFC, as observer, made a refuelling stop at Exmouth Gulf. The Catalina was stripped of armour and had extra fuel tanks fitted for its long flight, which used the US Navy flying boat base at Crawley Bay on the Swan River as its Perth base. The flight was operated as follows:

16-17 December 1942	
China Bay - Exmouth Gulf	26 hrs 15 mins
18 December 1942	
Exmouth Gulf - Perth	5 hrs 46 mins
21 December 1942	
Perth - Exmouth Gulf	5 hrs 22 mins
22 December 1942	
Exmouth Gulf - Cocos Islands	10 hrs 3 mins
22-23 December 1942	
Cocos Islands - China Bay	15 hrs 22 mins

A few weeks later a second crossing was made, this time in Y-49, to carry two crewmembers, 882 pounds of spare parts and 66 pounds of mail for the Royal Navy submarine HMS *Trusty* then at the Exmouth Gulf US Navy submarine base. Y-49 left China Bay on 14 January 1943, flying non-stop to and from Exmouth, before returning to China Bay on 20 January.

On 28 April 1943, Lieutenant Reijnierse made another crossing. He was sent to Perth to collect Vice Admiral Helfrich and his aide, Lieutenant von Fritjtag Drable, who were to make an inspection tour of military installations in Ceylon and India. Y-57 refuelled at Exmouth southbound to Perth, alighting at the Crawley Bay US Navy base on the Swan River. After a stay in Perth, the Catalina departed the Swan River on the return flight at 0301 on 20 May. They flew non-stop to China Bay in a flight time of 28 hours 31 minutes. At take-off the Catalina carried 1,804 gallons of fuel, of which only 154 gallons remained after mooring; the total distance covered was 3,634 miles with a ground speed average of 129 knots, and fuel consumption averaged 58 gallons per hour.

There was no direct connection between these MLD flights and subsequent RAF survey flights from Ceylon to Perth a few months later in preparation for the Qantas service. However, the gallant Dutch crews' experiences must have been of great benefit in the planning of the RAF trial flights.

Approval for a Qantas Indian Ocean Service

While in London from April 1943, Hudson Fysh had meetings with the Air Ministry, British government officials and BOAC senior management, lobbying for QEA's role in a new Indian Ocean service. In his book *Qantas at War*, Fysh describes his disappointment at finding the pre-war status of civil airlines had been trampled by RAF Transport Command which, he was told, planned to operate an Indian Ocean route as a purely military operation. His impression was that at that stage of the war, the Air Ministry was rightly focussed on military aspects to the almost complete exclusion of the part civil airlines could play in the war effort. He wrote an article *The Empire Wants an Air Plan Now* which *The Times* declined to publish. However, it was run in *The Daily Mail*, which led to questions being asked in The House of Commons.

Unknown to Fysh because of wartime security, within the Air Ministry there had been action on his Indian Ocean proposal and objections from RAF Transport Command were resisted. In fact, as early as March of that year, instructions had gone to Air Command South East Asia's No. 222 Group headquarters in Ceylon. These detailed a series of trial flights to be undertaken between Ceylon and Perth in preparation for a wartime secret service operated by BOAC. The RAF trial flights were to be carried out using two BOAC Catalina aircraft with armament removed and extra fuel tanks, which were being despatched to Ceylon for that purpose.

Finally in early June 1943, Fysh and BOAC officials were summoned to a meeting at the Air Ministry, at which the Indian Ocean plan was finalised. A non-stop Ceylon-Perth route was to be handed over to BOAC, which nominated Qantas Empire Airways as its agent to actually operate the service. RAF route survey flights from Ceylon were about to commence using two modified long-range BOAC Catalinas. When completed, the route would be handed over to Qantas. In his book *Qantas at War*, Hudson Fysh set out the broad financial details:

> The agreements for the service were regarded entirely as wartime expedients, in no way affecting the renewal of the QEA overseas service after the war. There was an agreement between the British Air Ministry and BOAC and in turn between the Corporation and QEA under which we agreed to operate the service on their behalf, receiving the nominal sum of £100 Sterling per annum as profit.
>
> All our capital expenditure and operating costs, including overhead, were recoverable from BOAC. The United Kingdom government had an agreement with the Australian government which, so far as I can ascertain from my records, provided that the UK government would supply the aircraft and bear the estimated cost of the operation up to £104,900 (Australian) per annum for a weekly frequency and £175,862 for a twice-weekly frequency. To assist in this expense the Australian government accepted an indebtedness of up to £52,000 sterling per annum plus payment for mails carried and certain responsibilities for ground facilities at the Australian end.

The original Air Ministry military plan was to operate RAF Catalinas from Trincomalee, a city on the southern coast of Ceylon to Exmouth Gulf, Western Australia. This gave the shortest crossing of the Indian Ocean, allowing a higher payload to be carried. The proposal was promptly dropped when the open sea conditions at both Trincomalee harbour and Exmouth Gulf were ruled as unacceptable. The overweight Catalinas would require calm water conditions for their long take-off runs. Therefore, Koggala Lake was chosen. It was an RAF Catalina and Sunderland base on a large lake on the southern coast of Ceylon, 80 miles from the capital city Colombo. At Perth, the smooth waters of the Swan River provided perfect water conditions, already used by US Navy "Black Cats" on Indian Ocean patrols.

The revised Koggala Lake-Perth route was 500 miles longer than via Exmouth Gulf, requiring maximum fuel to be carried, and reducing the payload to a mere 1,000 pounds. This comprised just three passengers, diplomatic dispatches and airmail.

RAF Survey Flights

No. 222 Group, RAF, at Colombo was tasked to operate a series of survey flights between Ceylon and Perth. The group's base was RAF Ratmalana, which had been Colombo's pre-war Ratmalana Airport. Across Ceylon the group had squadrons equipped with Beaufighters, Wellingtons, Spitfires, Sunderlands and Catalinas. Two Catalina squadrons were based at RAF Koggala on the banks of Koggala Lake on the far south coast of Ceylon. Today the site is Koggala Sri Lanka Air Force Base.

Because operational RAF Catalinas were in full military configuration with heavy fittings such as a nose gun turret and crew position armour plate, the Air Ministry sent two lightened BOAC Catalinas to Ceylon for the survey flights. The aircraft were G-AGFL and G-AGFM, both RAF machines transferred to BOAC for essential wartime routes to Lisbon and Africa. During civil certification inspections in October 1942 at BOAC's flying boat maintenance base at Hythe near Southampton, all military equipment was removed to save weight and rudimentary passenger seating was installed.

In March 1943 BOAC was instructed to withdraw the pair from scheduled services and send them to Hythe for inspections and modifications prior to long ferry flights to Ceylon by BOAC crews. They were handed over to No. 222 Group at Ceylon during April: G-AGFL reached Koggala on 21 April 1943, followed by G-AGFM four days later.

There were two resident RAF Catalina squadrons at Koggala at that time, No. 205 Squadron and No. 413 (Canadian) Squadron. The latter was selected to conduct the survey flights to Australia. The commanding officer was Canadian Wing Commander JC Scott, DSO.

The first survey flight took place on 3 May 1943. The Catalina, commanded by Scott, departed Koggala at 0846 and alighted on the Swan River at Perth at 0805 the next morning. The next southbound flight departed nine days later under the command of Wing Commander MD Thunder, RAF, and reached Perth on 13 May.

RAAF Western Area Headquarters senior navigation officer Flight Lieutenant Jim Cowan remembers their visit to Perth:

> Thunder, an Englishman who was full of thunder, had a little red moustache and was Wing Commander to the bootstraps. His navigator was a very quiet Canadian pilot officer. After a couple of weeks of high living in Perth, they set off for Ceylon early one morning. When it got dark that evening, the navigator took his first astro-fix and found that the intercepts he plotted were so long that he took another shot, to make sure. He was dismayed to find that he was about 600 miles from where he thought he was, and they were slowly flying further off track. There was no alternative but to turn around and come back to Perth after 24 hours in the air.
>
> I received a signal advising that the aircraft was returning and the captain wanted the compass swung, which was my job as Area Navigation Officer. On their return Wing Commander Thunder strode into my office and chucked the navigation logs on to my desk, requesting I look through them as well as having the compass swung. I examined the nav. log. In the column used to enter magnetic variation the navigator had entered only a stroke. An interesting feature of that route is that you always flew along one isogonal line which was the same all the way, at that time four degrees west. The navigator had flown all the way from Ceylon to Perth without changing variation, so he had plotted the homeward course also applying no variation. He was four degrees off course from the start. As they went further west, variation changed more and more until, by dusk, he was fifteen degrees off course. When they turned back. He flew almost the same path, so the error slowly cancelled out, and he hit Perth without difficulty. I telephoned him at his Perth hotel and asked him what variation he had applied. There was a deathly hush, and finally he said "Oh, my God!"

Five more Ceylon-Perth and return survey flights were flown by RAF crews. As these trial flights progressed, word came from the Air Ministry in London that Qantas Empire Airways would be operating a weekly service Koggala-Perth from July. These two BOAC Catalinas G-AGFL and G-AGFM would be the first of five transferred to the Australian airline.

QEA Western Operations Division Formed

Before confirmation came from the Air Ministry in London, Qantas commenced preparations to establish a new base in Perth should the proposed Indian Ocean service be approved. The airline's Long Range Operations Division formed in 1940 to carry out the RAAF Catalina deliveries was reactivated in April 1943 and renamed the Western Operations Division, to be based in Perth. Senior Captain WH "Bill" Crowther was appointed as the division's manager and given overall responsibility for the planned Indian Ocean operation. He immediately flew from Sydney to Perth on an Australian National Airways DC-2 service to commence his new and daunting task. Perth was new territory for Qantas.

A site for a Catalina base was selected on the Swan River, an undeveloped area of land at the river's edge in the Perth suburb of Nedlands. The location was conveniently close to the city and just a few miles from the existing US Navy Catalina base at Crawley Bay. Crowther requested experienced QEA Engineering Officer Norman Roberts to be in charge of engineering. Roberts had been a flight engineer during the Qantas delivery flights of RAAF

The Hotel Adelphi, Perth, was the temporary home for many Qantas aircrew sent to Perth for the secret Catalina operation. (JS Battye Library)

Catalinas and more recently was Senior Station Engineer at Darwin during the many Japanese bombing raids from February 1942. Despite the devastation, he had kept Qantas facilities at the civil airfield and flying boat base operational.

Intriguingly, as early as April 1943, QEA Empire flying boat First Officer HT Howse had been sent to Perth. That month in reply to a letter from the Department of Civil Aviation concerning his Aeronca Chief VH-ACH stored at Brisbane for the duration of the war, Howse gave his address as the Hotel Adelphi, Perth. He was not to receive a Catalina conversion course at RAAF Rathmines until November 1943, so it is assumed he was in Perth to assist Crowther in the early Perth days. Following his Catalina training course Howse was listed as a supernumerary crewman on QEA Empire service TN120 Sydney-Brisbane-Townsville on 6 December 1943, probably while returning home to Queensland on leave. He returned to Perth as a Catalina first officer, and was later promoted to a Liberator captain, still while quoting the Hotel Adelphi as his address in 1945. HT "Hoddy" Howse went on to enjoy a long Qantas postwar career on Constellations and Boeing 707s.

When Hudson Fysh cabled Crowther from London in early June to advise him that Qantas was approved to operate the service, work to establish the new Perth base was under way. Crowther had rented office space in Yorkshire House, St Georges Terrace, in the city, which became the administrative address for the QEA Western Operations Division.

Norm Roberts arrived in Perth in June 1943, accompanied by QEA Engineering Manager Arthur Baird to assist with the initial setting up. Experienced company engineer Colin C Sigley was appointed as the Nedlands Senior Engineer and maintenance ground staff were engaged.

With land clearing under way, Crowther and Roberts organised the necessary support facilities for maintenance, ground handling and refuelling, as well as a jetty and administration offices. A Ford V8 utility was purchased and kept busy transporting supplies of ropes and marine gear needed to handle the flying boats. With no storage yet at the base site, they were grateful to gain the use of four lockable dressing cubicles at the nearby Nedlands Swimming Baths.

Crowther arranged to be carried as a supernumerary crewmember from Perth to Koggala Lake on the return leg of one of the early RAF trial flights. He stayed on in Ceylon to make arrangements with RAF headquarters and local authorities for the handling of the Qantas flights. This included all aspects of civil airline passenger handling, including the ground transport for mail and passengers between Koggala Lake and Colombo. Following RAF survey flights carried more QEA aircrew for familiarisation, while the final return flights also carried QEA administrative personnel to RAF Koggala to set up airline facilities.

Meanwhile in Perth, such was the haste under strict wartime secrecy, to establish the Nedlands flying boat base, that there was little initial communication with support services. The Shell Oil Company held important Qantas contracts to provide refuelling services for its aircraft at Sydney and throughout Queensland. Shell Aviation's West Australian manager Dudley E Barker recalls a chance remark in June 1943 in his Perth office that senior Qantas people had just arrived in Perth. He set off around town making enquiries and found them in the lounge of The Esplanade Hotel – a party of Qantas "royalty" including no less than Operations Manager Captain Lester Brain and Captains Bill Crowther and Russell Tapp, all of whom he knew personally from his previous

Qantas business dealings. Barker remembers a convivial gathering at which he was told nothing of what was afoot, but from the guarded conversations he gathered he should be prepared. Shell was soon engaged to provide fuel and oil services for the secret Nedlands base. For the rest of the wartime QEA Perth operations, Shell and Baker personally enjoyed a close working relationship with the key men of the Qantas Western Operations Division.

At first Crowther was disappointed that the hoped-for Liberators had been refused. The Catalinas would have to fly a much longer ocean route between Ceylon and Perth, rather than a shorter landplane crossing to the RAAF airfield at Exmouth Gulf. At Perth they would use the wide sheltered waters of the Swan River, which ran past the city through southern suburbs to the port at Fremantle. He had been assessing water conditions and weather forecasts along the route including upper winds, which, together with then-accepted Pratt & Whitney R-1830 engine operation, made the long route marginal. He had cabled Fysh stating his serious doubts that Catalinas could operate a weekly schedule. When Qantas took over the first two Catalinas Crowther instituted a rigorous airframe weight saving campaign, to the extent of filing down protruding boltheads and fittings. After much experimenting on the early Qantas services, a revised engine operation plan gave the necessary endurance to allow scheduled services with few weather delays.

Engineer Norm Roberts recalled this period:

> Most difficult and destitute circumstances confronted our aim to establish the world's longest regular air service ever then attempted. Difficult because we had to start with no facilities for handling, servicing, fuelling or despatching our aircraft. Destitute because we did not have any servicing equipment or tools, nor any spares except for fifty sparkplugs I had collected in Melbourne on my way from Sydney to Perth.
>
> Further destitute because the stubborn opposition to the venture by DCA made it

A welcome view of the Swan River from a QEA Catalina inbound from the Indian Ocean. The Nedlands base is near the swimming baths structure in the water this side of the Pelican Point peninsula and the US Navy base hangars can be seen on the other side. (Neil Follett Collection)

> necessary for us to proceed without at least the primary aids and support that should have come from that quarter. It was a Qantas team effort and our determination to succeed helped us ignore many difficulties we had to overcome during those early months.

Roberts' criticism of the Department of Civil Aviation was understandable but perhaps a little unfair. The department found itself thrust into a supervisory role of an operation over which it had little control. The British Government had appointed BOAC to operate the Indian Ocean route with QEA acting as their agent – but with high-level British and Australian government diplomatic involvement under wartime security and secrecy. The DCA was required to supply airport facilities for Australian civil airline operations, but the sparse early information received by the department made it difficult to approve expenditure on what seemed to be a wartime military operation.

The DCA Assistant Director-General, Captain Edgar C Johnston, flew from Melbourne to Perth to personally review the Nedlands site. Departmental senior management was already troubled by wartime expediencies in which Australian civil airline pilots were flying military transport aircraft to New Guinea in support of the Allied New Guinea campaign - but this Indian Ocean service brought new complexities. Johnston's talks with Captain Crowther revealed that Qantas intended to operate with fuel loads well above civil certification limits. Because the Catalinas were British registered machines with British Certificates of Airworthiness (CofAs), DCA's role was uncertain. Overall, from the sparse official documentation that has survived, it seems that DCA adopted a "hands-off" attitude to the operational aspects, deeming it a wartime military operation. The department limited its role to aircraft inspection reports and overseeing the validation periods of each aircraft's British annual CofA renewal.

The main point of contention with the DCA was its failure to promptly arrange essential marine airport facilities such as a sealed slipway and hangar to allow the Catalinas to be pulled up on to land for maintenance and refuelling. The department's obligations to provide airport facilities for civil airline services seem to have been lost in the confusion over responsibilities in relation to this new Qantas quasi-military wartime operation. After a few months DCA sent instructions to the Commonwealth Department of Works to build a slipway at Nedlands, but that exercise caused further angst, as described later.

Roberts directed his aircraft airworthiness correspondence to the Australian DCA headquarters in Melbourne, which in turn forwarded some issues to the British authorities for response. DCA usually approved routine requests to extend Catalina annual CofA validation periods to allow the aircraft to fly additional services before being taken out of service for overhaul. DCA would send an aircraft inspector to Nedlands to report on the airframe and engines condition. An example was in July 1944 when Roberts requested two extensions:

- G-AGID CofA due to expire 20 July 1944. DCA extended the expiry to 1 September 1944
- G-AGIE CofA due to expire 21 July 1944. DCA extended the expiry to 17 August 1944

Later that year DCA did refuse Roberts' request for another six-week extension for G-AGFM, following an earlier six-week extension until 11 December 1944. The department's reply stated that based on an inspection of the aircraft and some pilot reports, it considered an overhaul was necessary. The engines had completed 620 hours, which was unsatisfactory for another return flight to Karachi.

Aircrew for the New Service

To maintain its reduced wartime Queensland civilian air services, military courier contracts and now the proposed Indian Ocean route, Qantas desperately needed additional aircrew. Other Australian airlines also needed pilots. DCA approached the Department of Air earlier that year explaining the urgent need for available RAAF pilots to be temporarily loaned to the civil airlines. After reviewing pilot numbers, the Air Board issued a notice inviting RAAF pilots to apply for transfers to civil airlines for an initial period of twelve months.

Among the first to be transferred was Flight Lieutenant Rex C Senior. He was flying Avro Ansons at No. 2 Air Navigation School at RAAF Nhill following a tour on Sunderlands in England with No. 10 Squadron. While at Nhill he completed an astronavigation course. He later wrote:

> Because of my experience on the Sunderland flying boats, I was assigned to Qantas Empire Airways which I joined on 9 March 1943. The next day Captain Bill Crowther flew us new pilots to Townsville where Qantas crews were flying Short Empire flying boats to transport military supplies to Port Moresby, Milne Bay, Groote Eylandt and Darwin, bringing back wounded to hospital. Our accommodation was in the US air force officers mess at the Townsville Hotel. The Japanese were attacking our destination ports virtually daily, while our civil aircraft were unarmed, so I quickly felt back in the war.

Early in June 1943 a small group of us in Townsville, including Captains Russell Tapp and Lew Ambrose (with whom I had flown often), John Solly and myself were informed we were to transfer to Perth to commence a new service. I was given leave with instructions to report to RAAF Rathmines on 23 June for conversion to the Catalina. After the two-day conversion course, on 25 June Captain Tapp and I flew by ANA to Perth where we rejoined our radio operator Glen Mumford and flight engineer Frank Furniss. Our accommodation was one of the premier hotels on the Perth Esplanade.

The first Catalina G-AGFM was flown down from Ceylon by RAF Squadron Leader Rumbold and crew and handed over to Qantas on 25 June. Our crew now commenced familiarisation and test flights with the intention of departing for Ceylon on 28 June. The weather was unsuitable on that day, so our departure was delayed until the following day. With Captain Ambrose and John Solly, we were the only two crews on the route until about September when more arrived in Perth.

Because of our secret operation, we were instructed to wear civilian clothes except while on duty. Following several weeks in a city hotel I obtained accommodation in a guest house near Kings Park and from there enjoyed the walk down to Captain Crowther's office in Yorkshire House, St Georges Terrace. On one such occasion I stopped to buy a packet of cigarettes from the park kiosk, only to be told in very harsh tones "Go and join up, there are no cigarettes for bludgers" while on another occasion I was the recipient of a white feather in an envelope.

Another RAAF recruit was Ashley Gay who had enlisted as an airframe fitter before being transferred to aircrew. In July 1943 he had just completed his pilot training at RAAF Point Cook on Airspeed Oxfords when his course was told the Air Board was calling for volunteers for secondment to commercial airlines. He was pleased to be allocated to Qantas and on 8 August commenced as a second officer on the Empire flying boat military courier service Sydney-Brisbane-Gladstone-Townsville. After being sent to RAAF Rathmines for Catalina training he was transferred to Western Operations Division, making his first training flight on the Swan River on 21 January 1944 in G-AGIE *Antares Star* with Captain Bert Ritchie. Second Officer Gay made his first Indian Ocean crossing Perth-Ceylon on 10 February in G-AGID *Rigel Star* under the command of Captain Frank Thomas. Gay recalled being presented with the tongue-in-cheek second officer's pin, crossed knives and forks to signify their unofficial role as flight stewards.

RAAF Western Area Headquarters Perth posted notices inviting applications from pilots with astronavigation training for twelve-month secondments to civil airlines. Sergeant Ivan Peirce was flying Avro Ansons from Geraldton on West Australian coastal patrols, but his training had included an astronavigation course at No. 2 Air Navigation School, Nhill. In March 1943 he applied and was transferred to RAAF Reserve as "Employed by Civil Airline" and told he was going to Qantas. Ivan later recalled:

> It was all so secret they wouldn't tell us anything. They just said, "You've got to volunteer for it, and you'll be going into operational areas." Well at that time, there'd been two years of war, and we hadn't seen an operational area, so we agreed to it. I did flying boat training in Sydney on the Qantas Empires, circuits at Rose Bay then went up to Townsville and Port Moresby and places like that for about two months.
>
> Sent over to Western Australia, Qantas picked you up at the airport and you're staying in the Esplanade Hotel in Perth and that was that. I had a couple of days' training on the Catalina on the Swan River and finally found out where I was going. You weren't even allowed to tell your family. We had to get a passport because we were civilians. We wore our air force uniforms, but with Qantas rank and Qantas wings. If captured by the enemy we were told to give our name, air force number and rank and nothing more. Secrecy was a really top priority. And I think that was the success of it too, because the

Catalina G-AGFL Vega Star at Nedlands, Perth, in the early months of the service before the boarding jetty had been completed. The Nedlands swimming baths are to the left. (Geoff Goodall Collection)

Japs would listen to every bit of radio chatter that was around, but they never picked us going across the Indian Ocean.

QEA pilots were arriving in Perth, having been sworn in as registered members of the RAAF Reserve at Rose Bay prior to their departure for the west. They were fitted out with RAAF uniforms and full flying kits at Bradfield Park Supply Depot, Sydney. Basic type-conversion to the PBY-5 Catalina was conducted by No. 3 Operational Training Unit at the Rathmines flying boat base. On arrival in Perth the early ones received Catalina training by Captain Crowther on the Swan River during the RAF survey flights.

Crowther was joined in Perth by fellow senior captains Russell Tapp and Lewis Ambrose to supervise the flying side of the operation. All were highly experienced Qantas men, having flown the Australia-Singapore service on Short Empire flying boats and the earlier DH.86 landplanes. As the influx of Qantas employees increased, finding them places to live was a problem. Accommodation in wartime Perth was scarce; there were very few places available for rent, and top-class city hotels gained some QEA long-term guests, including many of the unmarried staff. Initially Western Operations personnel were not permitted to have their families with them in Perth. After six months, when the success of the operation was assured, families were permitted to move to Perth, provided they could obtain the necessary wartime travel warrants to book seats on the three-day train trip from the eastern states, not easy for non-military personnel. Apart from aircrew, some of whom preferred private board with Perth families, the majority of the additional staff recruited were Perth locals, who could live at home.

The Nedlands base was made ready to commence virtually a new airline operation, 2,000 miles from the Qantas head office in Sydney, and independent from other Qantas services, in wartime and in secrecy.

Chapter 3 Catalinas: Double Sunrise Service

The Catalinas

Large numbers of Consolidated Catalinas were being supplied to Great Britain from the United States under the terms of the wartime Lend-Lease Act. A small number had been released from the RAF to BOAC to maintain civilian services mainly to Africa. When Qantas received British Air Ministry approval to operate the Indian Ocean service, it was agreed that four Catalinas would be supplied by Britain to maintain a weekly service, allowing for airframe and engine maintenance. They would be flying boat models without the weight of the retractable land undercarriage of amphibian model Catalinas. Later, in May 1944, a fifth RAF Catalina would be transferred to BOAC for handover to Qantas to increase capacity on the Indian Ocean service.

The first two BOAC Catalinas for the Indian Ocean service, G-AGFL and G-AGFM, had been sent to Ceylon in March 1943 for the planned RAF Ceylon-Perth survey flights. The next two for the Indian Ocean service were recently delivered PBY-5 flying boats allocated to BOAC, to be passed on to Qantas. They were ferried to BOAC's Hythe flying boat base on the English south coast in July 1943 for civil conversion and installation of six additional fuel tanks near the flight engineer's station. After British civil certification as G-AGID and G-AGIE, both set off during August for the long ferry flight by BOAC crews to Ceylon. G-AGID reached Ceylon on 26 August where it was handed over to Qantas at Koggala Lake. Three days later its delivery to Perth continued under the command of Captain Ambrose as a special flight making a brief enroute stop at the Cocos Islands to drop off two essential meteorological observers, as described later. This Catalina's first scheduled service was 2Q8, departing Perth for Koggala on 8 September 1943 under the

Raising the bow anchor of a Catalina at Nedlands prior to departure for another Indian Ocean crossing. This image of G-AGFM Altair Star shows the name with the "star" symbol and "QANTAS EMPIRE AIRWAYS LTD" as painted in black on the forward section of the Catalinas. (Qantas)

G-AGFM "on the step" taking off at Lake Koggala, Ceylon, showing its camouflage and fuselage roundels. (Imperial War Museum)

command of Captain OFY Thomas. Qantas flight numbers are explained at the end of this section.

G-AGIE departed Hythe on 17 August 1943 but was forced to return from Lisbon due to low oil pressure in one engine. After repairs by BOAC at Hythe it was diverted again due to a combination of autopilot failure and an elevator control problem, having these rectified at the BOAC flying boat base at Foynes, Ireland. Finally departing Foynes

Civil Reg.	QEA Star Name	Fin No.	Received QEA	C/n (Hull Number)	US Navy type	US Navy serial	RAF type	RAF serial
G-AGFL	*Vega*	1	12.7.43	808	-	-	Catalina 1B	FP221
G-AGFM	*Altair*	2	25.6.43	831	-	-	Catalina 1B	FP244
G-AGID	*Rigel*	3	26.8.43	1109	PBY-5	Bu08215	Catalina IVA	JX575
G-AGIE	*Antares*	4	13.9.43	1111	PBY-5	Bu08217	Catalina IVA	JX577
G-AGKS	*Spica*	5	16.5.44	28022	PB2B-1	-	Catalina IVB	JX287

on 3 September, it routed via Lisbon, Gibraltar, Cairo, Basra, Karachi and Madras, before reaching Koggala Lake on 13 September for handover to Qantas. G-AGIE's first service was 1Q12 Koggala-Perth on 27-28 September 1943.

The Catalinas were delivered in RAF camouflage. The colours were Extra Dark Sea Grey and Dark Slate Grey on the upper surfaces which gave them an overall sea green appearance. The under surfaces were pale blue Sky. They retained their RAF serials, roundels and fin flashes. The roundels were RAF South East Asia Command blue and white only, with the red centre removed to avoid confusion with the Japanese aircraft *hinomaru* red disc. The British civil registrations were painted over the camouflage on the aft fuselage sides, under the wings and also on the top of the wings of some aircraft. The most prominent marking was a large "fleet number" 1 to 5, painted in white on each side of the fin by BOAC prior to delivery. The purpose of these numbers remains obscure, as no reference to their allocation has been found in airline records.

Captain Crowther suggested the aircraft should be given individual names from the stars by which they were navigated. After each aircraft arrived in Perth, both sides of their bows had "QANTAS EMPIRE AIRWAYS LTD" painted in black over the camouflage with the star name and a six-sided star emblem. The first official mention of the "star" names was in a Western Operations Division report by Crowther to Qantas Sydney management in September 1943. Qantas aircrew recall that they always referred to the Catalinas by their civil registrations or star name, rather than their number.

The following summaries of the earlier careers of these aircraft are thanks to the research of Catalina historian David Legg:

FP221 (G-AGFL)

25.7.42	Accepted by RAF
10.8.42	Delivered ex San Diego, NAS Elizabeth City, North Carolina, to Gander Lake, Newfoundland
1.10.42	Allocated to BOAC on loan
8-9.10.42	Gander Lake-Lough Erne, Northern Island
10.10.42	Greenock, Scotland for modifications by Scottish Aviation Ltd
27.10.42	Registered G-AGFL to BOAC
31.10.42	CofA issued
3.11.42	Commenced BOAC service on Poole-Lagos route
17-25.4.43	Ferried Poole to Ceylon by BOAC crew, handed over to QEA

FP244 (G-AGFM)

6.8.42	Accepted by RAF
10.9.42	Delivered ex San Diego, NAS Elizabeth City, to Gander Lake
1.10.42	Allocated to BOAC on loan
8-9.10.42	Gander Lake-Lough Erne, Northern Island
10.10.42	Greenock, Scotland for modifications by Scottish Aviation Ltd
27.10.42	Registered G-AGFM to BOAC
31.10.42	CofA issued
11.42	Commenced BOAC service on Poole-Lagos route
11-20.4.43	Ferried Poole to Ceylon by BOAC crew, handed over to QEA

JX575 (G-AGID)

16.12.42	Accepted by US Navy as BuNo 08215.
31.12.42	Diverted to RAF
9.1.43	Delivered ex San Diego to US Navy Fleet Air Wing CFAW-14, Elizabeth City, North Carolina
9.1.43	Taken on RAF charge as JX575
10.5.43	Delivered by RAF to Largs, Scotland
7.6.43	Greenock, Scotland for modifications by Scottish Aviation Ltd
5.7.43	Registered G-AGID to BOAC
13.7.43	Greenock-Hythe on delivery to BOAC
21.7.43	CofA issued
15-26.8.43	Ferried Hythe to Ceylon by BOAC crew, handed over to QEA

JX577 (G-AGIE)	
17.12.42	Accepted by US Navy as BuNo 08217.
31.12.42	Diverted to RAF
12.1.43	Delivered ex San Diego to US Navy Fleet Air Wing CFAW-14, Elizabeth City, North Carolina
12.1.43	Taken on RAF charge as JX577
14-15.5.43	Delivered by RAF via Gander Lake to Largs, Scotland
18.5.43	Scottish Aviation Ltd, Largs for modifications
5.7.43	Registered G-AGIE to BOAC
13.7.43	Greenock-Hythe on delivery to BOAC
21.7.43	CofA issued
17.8-13.9.43	Ferried Hythe to Ceylon by BOAC crew, handed over to QEA

JX287 (G-AGKS)	
17.1.43-2.2.43	Delivered by RAF Elizabeth City to Bermuda, Gibraltar
2.43	RAF New Camp, Gibraltar, later 45 (Atlantic Ferry) Group
29.1.44	Allocated to BOAC on loan
1.2.44	Damaged Gibraltar, wingtip struck moored No. 202 Squadron Catalina
2-3.3.44	Gibraltar-Largs. Scotland
16.3.44	Delivered to BOAC at Hythe
6.4.44	Registered G-AGKS to BOAC
2.5.44	CofA issued
7-16.5.44	Ferried Hythe to Koggala by BOAC crew, handed over to QEA

The last aircraft G-AGKS *Spica Star* was the "problem child", which was to cause Qantas Western Operations Division many troubles during its relatively short career. It differed from the others, being a late production PB2B-1 built under licence by Boeing Aircraft of Canada at Vancouver, whereas all the others were Consolidated-built at San Diego. Although construction was identical, the Boeing-built aircraft incorporated various changes and refinements. The cockpit layout had differences, including the float-retraction switch being fitted where the other Catalinas had starter-motor switches - leading to several embarrassing episodes when captains raised the wing floats at the jetty as onlookers expectantly waited for engine start-up!

A more serious problem of fuel leaks plagued *Spica Star* from the time it was delivered. The Davis wing, which characterised many Consolidated aircraft, was a high aspect ratio design: the primary

The troublesome fifth Catalina G-AGKS Spica Star at the Nedlands jetty, circa 1944. (Qantas)

structure being a wide box spar into which the integral fuel tanks were built. This type of wing is subject to considerable flexing in flight, and for this reason Consolidated had riveted neoprene gasket strips between all the mating surfaces in the wing tank section. This method of sealing the tanks was entirely successful, but time-consuming to construct. Boeing bypassed this method by assembling the structure without gaskets but sealed the interior joining sections with synthetic filling putty. This method is used today with absolute reliability but was not dependable at that time.

Qantas repaired the fuel leaks as each developed, but the problem continued. In November 1944 the troublesome Catalina was flown from Perth to Rose Bay to have the wing fuel tanks completely resealed. This involved considerable effort, but the result was successful. It was test flown at Rose Bay on 23 December by Captain GU "Scotty" Allan, who had returned to Qantas after leaving for RAAF service at the beginning of the war. G-AGKS *Spica Star* was ferried back to Perth and on 4 January 1945 departed on a Perth-Karachi return service. On its return to Nedlands on 10 January, it was retired as part of reduced Catalina schedules as Qantas Liberators took over the brunt of the Indian Ocean route.

Early estimates calculated that the Catalinas would not be able to maintain altitude on one engine until eight hours after take-off, when sufficient fuel had been consumed to reduce weight. When negotiating for the supply of Catalinas, Hudson

A view of Lake Koggala from a departing Qantas Catalina. Underneath the wing is one of two fuel dump pipes fitted to each aircraft. (EH Neal)

Fysh had insisted firmly, against British opposition, that they be delivered with a fuel-dumping system installed. This would allow a quick reduction of weight in the event of engine failure, to allow the aircraft to continue flying on one engine although, of course, with a much-reduced range. Fysh argued that the weight penalty of the modification's plumbing was compensated by the safety margin provided. Despite assurances this would be carried out, the Catalinas were delivered without fuel-dumping fittings. In order for the operation to commence as soon as possible, a number of Indian Ocean crossings were made without the ability to dump fuel in an emergency. When the fuel dump fittings arrived in Australia, each Catalina was taken out of service for an overhaul while these modifications were installed. The wisdom of Fysh's perseverance was justified by subsequent events.

Each Catalina was progressively withdrawn and sent to Sydney for maintenance at the Qantas workshops at the Rose Bay flying boat base in Sydney Harbour. The work included renewal of the annual British Certificate of Airworthiness and modifications. The long-awaited fuel dump systems were installed and the extra fuel tanks moved forward in the cabin to improve the centre of gravity position. Previously, in the early stages of the long flight, with the trim wound fully forward, forward pressure on the control column was required to maintain level flight. Each of these extra tanks held 66 imperial gallons of aviation fuel. Normal Catalina tankage was 1,460 imperial gallons; these extra tanks allowed 1,988 gallons to be carried.

First to go to Sydney was G-AGFM *Altair Star*, which flew from Perth on 24-25 September 1943, in a flight time of 15 hours 21 minutes. Captain Tapp test flew G-AGFM at Rose Bay on 19 October 1943 for 1 hour 40 minutes during which the fuel jettisoning system was checked, prior to departing on the long ferry flight back to Perth. *Altair Star* was only briefly on the water at Nedlands before being dispatched on 21 October to operate service 2Q14 to Ceylon. Fuel was later dumped operationally on six occasions.

Next to go to Sydney for modifications was G-AGID *Rigel Star* in late October 1943. On completion at Rose Bay Captain Bert Hussey took it on an acceptance test flight on 22 November before ferrying the Catalina back to Perth on 24-25 November in a 16-hour 43-minute flight. Two days later *Rigel Star* operated Service 2Q21 from Perth to Ceylon and Karachi.

Qantas Flight Numbers

All scheduled airlines have separate flight numbers for each service they operate. When Qantas commenced international services in 1935 in conjunction with Imperial Airways Limited, the IAL "Service Code" system was adopted for uniformity. This designated the route and direction. For example, "SE56" was the 56th Sydney eastbound flying boat service. Imperial Airways became BOAC and from 1942 when the war cut Qantas services off from those of BOAC, Qantas continued to use the same system but with new codes for its revised routes.

During 1943 BOAC implemented a new system of Service Codes that applied across associated airlines. The code comprised a designation number, a letter indicating the airline operator (Q for Qantas) and a number indicating the iteration of that service. For example, 2Q18 was the 18th "2Q" service: Catalina/Qantas/Perth-Ceylon-Karachi.

The following Service Codes relate to Qantas services described in this account:

1Q	Koggala-Perth	QEA Catalina service July 1943-July 1945
2Q	Perth-Koggala	Extended to Karachi from November 1943
3Q	Colombo-Perth	QEA Liberator service
4Q	Perth-Colombo	June 1944 to November 1945
5Q	Hurn-Sydney	BOAC/QEA Lancastrian service via Ceylon
6Q	Sydney-Hurn	June 1945 to April 1946
7Q	Hurn/Heathrow-Sydney	BOAC/QEA Lancastrian service via Singapore
8Q	Sydney-Hurn/Heathrow	April 1946 to November 1947
9Q	Negombo-Perth-Sydney	QEA Liberator service December 1945 to April 1946
10Q	Sydney-Perth-Negombo	
11Q	Singapore-Sydney	QEA Liberator service via Darwin
12Q	Sydney-Singapore	April 1946 to September 1946

Qantas Service Begins

The final RAF Survey flight overnight 24-25 June 1943 under the command of Squadron Leader Rumbold also served as the delivery flight of G-AGFM to Qantas. After alighting on the Swan River in Perth, it was handed over to Captain Crowther representing Qantas Empire Airways. The Catalina was used for crew training before a return to Ceylon three days later. This would be a Qantas positioning flight to enable the first scheduled Service 1Q1 to operate Ceylon-Perth in early July.

The Qantas crew for the company's first Indian Ocean crossing comprised Senior Route Captain Russell B Tapp, First Officer Rex C Senior, Radio Operator Glen W Mumford, Flight Engineer Frank S Furniss and Navigation Officer Captain WH Crowther. A consignment of mail was carried, along with the returning RAF crew and QEA ground engineer CW South who was taking up his appointment as station engineer at Ceylon. An attempted take-off from the Swan River on 28 June was abandoned because of rough water conditions, the flight was postponed until the following

The Qantas crew at Lake Koggala prior to the inaugural southbound service 1Q1 to Perth in G-AGFM on 10 July 1943. From left to right Captain RB Tapp, First Officer RC Senior, Radio Operator GW Mumford, Flight Engineer FS Furniss and Navigation Officer Captain WH Crowther. (Civil Aviation Historical Society)

morning. G-AGFM departed Perth in the pre-dawn darkness at 0430 on 29 June 1943 and reached Koggala Lake at 0855 the next morning. This flight was recorded in the aircraft's journey logbook as "Indian Ocean Air Mail" and could be recognised as the commencement of the Qantas service. The generally quoted 10 July 1943 commencement date was in fact the first southbound Ceylon-Perth Service 1Q1.

First Officer Rex Senior described that first Qantas flight across the Indian Ocean:

> We departed from Nedlands at 0430 in the morning with our Catalina *Altair Star* nearly four tons overload and slowly taxied out from the primitive loading ramp. We turned into wind facing down the Swan River and, with the controls held as fully back as possible, the throttles were opened to full power. Take off in this condition required an extremely long run, with the fuel weight pushing our hull deep into the water. As we gathered speed the wing floats were retracted and the aircraft gradually lifted up on to the planing hull, then with a clear run of nearly four miles down to the Fremantle Bridge, we slowly gained enough lift for the Catalina to become airborne and commence our cautious climb out over Rottnest Island where we began a

G-AGID moored on the Swan River at Nedlands, Perth. (Neil Follett collection)

gradual turn to the right to take up our heading along the coast towards Exmouth Gulf.

In addition to our Qantas crew we carried as passengers the returning RAF crew members who provided help and advice, with their navigation officer of particular assistance.

We were very alert and conscious of the route we were initiating, and I think a little apprehensive. Cruising up the coast at 1,000 feet not far offshore that morning gave us the opportunity to familiarise ourselves with the largely undeveloped coastline. With autopilot engaged we carefully adjusted the trim and power to maintain our 99 knot indicated airspeed. Both Russell Tapp and I were too conscious of the significance of our flight to take time off duty. It was fairly late in the day when we approached Exmouth Gulf, our turning point for the ocean crossing and began the long section of the track to Ceylon. I acted as navigator for 7 hours, taking turns with Russell Tapp and the RAF navigator.

When dawn broke we were still 400 miles from our destination and in Japanese patrolled airspace, our lookout became most alert. Sextant sightings late in the night had shown we were well on track and after a while land began to appear on the horizon. At 0855 Perth time the skipper landed on Koggala Lake's smooth waters, and we taxied to the RAF station. This initial "Double Sunrise" flight took 28 hours and 10 minutes.

The first eastbound Qantas service took place ten days later when G-AGFM *Altair Star* took off from Koggala Lake on 10 July 1943 for Perth, operating "Indian Ocean Flight 1Q1". It was operated by the same crew under Captain Tapp, joined by Captain Crowther who had been having talks with the RAF in Ceylon and a loaned RAF navigator. The payload comprised 52 pounds of diplomatic and armed forces mail – one high priority passenger was to have been carried on this first scheduled crossing, but at the last moment failed to load. The flight was eventful because, some hours after departure, all the crew, except Tapp and Crowther, were struck down by food poisoning and became extremely ill. With the Catalina flying on autopilot Tapp and Crowther operated all the crew positions between them. An astronavigation fix at midnight revealed their groundspeed had dropped from 130 knots to 80 knots. They climbed 2,000 feet in an attempt to locate more favourable winds, while considering a diversion direct to the North Western Australian coastline for a forced landing. Fortunately, the wind improved at the higher altitude, and they continued to Perth where they alighted after a flight time of 28 hours 9 minutes. When the Catalina was moored at Nedlands, it was found to have 400 gallons of fuel remaining in the tanks, a fairly healthy margin. In the aircraft's journey logbook Captain Tapp wrote next to the Perth arrival time "1st Service Indian Ocean 1Q1".

Lost to history is the inglorious fact that the scheduled date for Service 1Q1 had in fact been 7 July 1943. On that morning Tapp departed Koggala Lake in G-AGFM. Three hours into the flight, the different loaned RAF navigator realised he had inadvertently left his two sextants behind in the RAF Operations Room safe. Without these essential navigation instruments, Captain Tapp had no choice but to turn around and return to Koggala. The navigator's name was not recorded, nor the captain's reaction when told of the complication. The Catalina had been airborne for seven hours and fourteen minutes when it alighted back at Koggala Lake on this aborted first scheduled service.

The same aircraft flew the first scheduled westbound service, 2Q1, leaving Perth on 22 July 1943, again under Captain Tapp's command, and reached Ceylon after a flight time of 28 hours 56 minutes. The second Catalina, G-AGFL, was having maintenance at Koggala and was scheduled to commence with Qantas on 13 July as service 1Q2 to Perth. There was a delay, and it did not get away until 21 July, reaching Perth the following afternoon. One flight per week in each direction was now operated.

Hudson Fysh, returning from his London visit, was a passenger on G-AGFL *Vega Star* from Ceylon to Perth on 30-31 August 1943 under the command of Captain Crowther. This flight recorded the longest crossing of the whole operation: 31 hours 45 minutes. Crowther's pilot logbook confirms 31 hours 45 minutes flying time, however a Western Operations Division report to head office listed the buoy-to-buoy time of 32 hours 9 minutes, which has sometimes been quoted as the flying time.

The Indian Ocean schedule was expanded to three flights per fortnight in each direction from 15 October 1943 when the third and fourth Catalinas entered service.

Operating Techniques

The Catalinas had a designed Maximum All-Up-Weight of 27,000 pounds. The British Certificates of Airworthiness for the Indian Ocean Catalinas increased the MAUW to 33,000 pounds. At QEA's instigation, BOAC requested the British Air Ministry to allow a further increase. On 9 January 1944 a cable from London to the Australian Department of Civil Aviation advised that that the British CofA MAUW for these specified Catalinas was amended to 35,000 pounds. This was based on RAF advice that its Ceylon-based Catalinas were departing at that weight for oceanic patrols. Notwithstanding this ruling Qantas had been operating up to a limit of 35,150 pounds for some months, the authority for which is lost to time.

Operating at that weight, the aircraft carried seven tons of fuel on take-off: a total overload of four tons! After early flights, Qantas adopted a special take-off technique: the control column being held central as full power was applied, then when effective control was gained, the pilot pulled the controls back as hard as possible, to bring the nose up and put the Catalina "on the step". This differed from the normal military procedure of holding the control column hard back from the beginning of the take-off run - but Qantas crews found their method reduced the take-off run by 10 to 12 seconds.

Navigator Jim Cowan summed up the prevailing attitudes:

> We knew that if we lost an engine in the first eight hours of the flight we wouldn't be able to maintain altitude and slowly but surely we would go down into the "drink". Fortunately, we rarely lost an engine during that first eight hours. The psychology of wartime flying is quite different from flying commercially as a profession, in that you don't really attach the same significance to taking risks or chances.

As a point of comparison, the DCA placed a MAUW of only 27,500 pounds on post-war civilianised Catalinas, including some operated from Sydney by Qantas, calculated on their ability to maintain altitude on a single engine.

Qantas aircrew followed an agreed company engine-operating plan, which had been developed in consultation between the captains and station engineer Roberts. Cruising air speed was 99 knots indicated, giving a True Air Speed (TAS) of 110 knots and rising to 115 knots as the fuel burn lightened the aircraft. Manifold pressure was adjusted to achieve a fuel burn of 22 gallons per hour for each engine. Senior First Officer EH Neal later described the issues involved:

Long range flights of this nature presented a number of in-flight operations problems:

1. The massive overload in relation to the aircraft's flying attitude.

2. The amount of fuel available, refuelling was not possible.

3. The unusually long distance.

4. A suitable airspeed was required in order that an out-of-trim flight attitude would be avoided, namely a tail down high drag situation.

5. Adverse winds over all or part of the route.

Broadly speaking, the problem could be stated in the following form:

(a) High airspeed = high engine power = high fuel consumption but low endurance. Lack of fuel would terminate the flight.

(b) Low airspeed = low engine power = low fuel consumption but high endurance. Lack of distance covered would terminate the flight.

Therefore, a compromise was adopted after studying the early RAF survey flights. The Qantas services were flown at 110-112 knots (TAS), then increased to 115 knots TAS. Higher air speeds could be flown depending on meteorological considerations.

On reaching cruising altitude, the power was adjusted for a TAS of 115 knots. As the fuel was consumed, the aircraft became lighter and increased its airspeed, so periodically the power was reduced to maintain the planned airspeed, thus reducing fuel consumption.

The following figures relate to the early and late stage of an actual flight from Perth to Ceylon:

- 0222 hrs GMT *Vega Star* left the buoy at Nedlands

- 0300 hrs GMT settled down on cruise, altitude 1,700 feet, 3,019 nautical miles from Koggala Lake, TAS 118 knots, Engine RPM 2,000, manifold pressure 30.5 inches, fuel consumption 76 gallons per hour, fuel remaining 1,902 gallons, and AUW 34,860 pounds.

- 24 hours later and 47 minutes away from the buoy on Koggala Lake, *Vega Star* was about to descend from an altitude of 2,000 ft. TAS 120 knots, RPM 1,750, manifold pressure 30.5 inches, fuel consumption 55 gallons per hour, fuel remaining 387 gallons and AUW 23,973 pounds.

At least a further seven hours of flying was available from the remaining quantity of fuel.

Due to the prevailing winds, in general Perth-Ceylon was flown at low altitude, usually 1,500 to 2,000 feet, while Ceylon-Perth was flown at 11,000 to 14,000 feet. At the higher altitudes the outside air temperatures would be minus sixteen degrees Celsius. With cabin heating and insulation removed, the crew wore RAAF issue thick flying suits and fur-lined flying boots. Passengers were warned not to touch the fuselage framework with bare hands because their fingers would stick to the metal. Oxygen was not carried because of the weight penalty of oxygen cylinders.

Various climb-cruise techniques were used to conserve fuel. The normal departure from Koggala southbound to Perth entailed cruising for the first three hours at 1,000 feet to take advantage of the prevailing northerly winds, then a gentle climb to 4,000 feet. After more fuel was burnt, a climb was made to 11,000 feet to break through the usual cloud cover for an astro-fix, then the aircraft settled down for the long cruise to Perth at the altitude giving the best wind conditions. Eastbound, climbs to a cruising altitude of 11-12,000 feet were made once the very active Inter-Tropical Front had been passed. The Inter-Tropical Front follows, but lags behind the sun's seasonal movements. The Qantas engine operation plan gave the Catalinas an endurance of 36 hours, equating to a range of approximately 4,600 miles subject to prevailing upper winds.

Catalina Senior First Officer Rex Senior:

> I flew predominantly with Captain Russell Tapp who excelled as a pilot. I was taught to carefully adjust the angle of attack of the wing, to where the slight decrease in air resistance made a difference to fuel consumption.

The flight engineer was responsible for monitoring the engines and fuel. The Catalina's Pratt & Whitney Twin Wasp engines burned fuel from the main wing tanks only. A few hours after take-off on each ocean crossing, the flight engineer, on instruction from the captain, would commence transferring fuel from the auxiliary tanks inside the fuselage to the wing tanks; this process continued for a number of hours.

Service 2Q97 Perth to Ceylon in G-AGIE *Antares Star* on 15 November 1944 was the first crossing for newly rated Flight Engineer John Maskiell, who was keen to impress. He had been one of the first ground maintenance staff employed at Nedlands. Five hours into the flight when he received the instruction from Captain MacMaster to commence fuel transfer, he replied that he had been doing so since soon after take-off from Perth: fuel he had diligently transferred (some 200 gallons) had resulted in the almost-full wing tanks overflowing through wing vents. Unfortunately, Maskiell's enthusiasm caused this service to be aborted: the fuel loss forced MacMaster to return to Nedlands. Before landing on the Swan River, an additional four tons of fuel had to be jettisoned to reduce the aircraft's weight to the maximum landing weight. There were no passengers, only mail and priority freight. Maskiell survived the dressing-down he received from MacMaster and Crowther and continued as a flight engineer until the end of the Catalina service. He continued with Qantas and later rose to chief engineer with Trans-Australia Airlines.

The radio was Morse code; there was no voice contact. The radio operator maintained a continuous listening watch for broadcast weather and operational messages. No acknowledgement transmission was sent from the aircraft. In fact, the radio transmitter power was turned off for most of each ocean crossing to eliminate any chance of a carrier wave being transmitted. Captain Ambrose wrote:

> Due to the requirement to maintain radio silence during all but the first and last stages of the flight, the aircraft radio served during a normal flight as an important listening post only. Qantas drew up a radio plan for normal and emergency use, in cooperation with the RAAF in Australia and the RAF in Ceylon. Before departure, each Catalina service was briefed in the official fashion with a complete list of stations, call signs and frequencies. The air force stations at each end maintained a continuous radio listening-watch on our frequency. It was agreed that any message for us would be transmitted during the first ten minutes after the hour GMT. As no answer could be expected from the aircraft, all messages were broadcast slowly and carefully to ensure our correct reception.
>
> In order to confirm weather forecasts, a special broadcast ... was made just before the aircraft reached the halfway mark, by the station of destination. This was repeated on several frequencies in order to ensure receipt. Enemy and other interference sometimes prevented our receiving this broadcast, but mostly it was received, decoded and put to good use. A further broadcast was made later in the flight of terminal conditions expected, but, as this was outside the zone of silence, it was answered.

Navigation

The original route for Perth departures was to fly north paralleling the coastline until Cape Inscription near Exmouth Gulf, then to follow a great circle track across the Indian Ocean to Ceylon. As the crews became comfortable with the conditions, the route changed to a great circle track from Rottnest Island, just off Perth, direct to Ceylon. On the return flights from Ceylon, all flights planned landfall at Exmouth Gulf to provide the safety of the coastline on the last leg into Perth. Departure times were adjusted to ensure each flight would transit Japanese patrolled territory during darkness.

The determining factor in the success or failure of the Indian Ocean service was navigation. Without precise navigation, regular services across the vast stretches of the Indian Ocean would not have been possible. The experience gained during the Pacific

Navigation Officer HJ Bartsch takes a celestial navigation fix at the Catalina's rear blister compartment. (Qantas)

Ocean deliveries of RAAF Catalinas had been invaluable, giving Qantas the confidence that it could operate the longer distances of the Indian Ocean.

Oceanic navigation was carried out solely by dead-reckoning and astronavigation. Navigation equipment comprised a Dalton Dead-Reckoning Plotting Calculator, sextants, accurate watch-chronometer, Norie's Nautical Tables, plotting charts and instruments. Two sextants were carried, both standard military issue Henry Hughes & Sons models. One was a Mk. IX type, which incorporated a manual averaging device, enabling snap astro sights to be made through gaps in cloud cover on the low altitude westbound flights. The second sextant was a Mk. IXA. It was fitted with a clockwork device which averaged the sights over a two-minute period: designed for accuracy, provided the star remained visible from the aircraft for the full two minutes, but otherwise useless. It had some value on the high-altitude eastbound flights.

The Catalinas carried fifteen flame floats, which could be dropped from the blisters at night to allow the navigator to take drift sightings. Each float weighed three pounds. For a short period, their supply was interrupted, and the replacement larger floats weighed eleven pounds each. This added weight forced a 120 pounds reduction in the payload that could be carried. Such were the tight weight margins.

Qantas Western Operations Division navigators had to identify the stars of both the Southern and Northern Hemispheres because they crossed the Equator. First Officer Rex Senior, who acted as navigator on early flights before RAAF navigators joined the crews, took sights as often as every half hour when required. He remembers becoming sufficiently experienced to identify a star if only seen briefly through a break in the clouds.

Navigationally, the three most important times of a flight were: just after dark, prior to dawn and at the time the midway weather forecast broadcast was received by radio. It was imperative for the navigator to get a reliable position fix before dawn; since much of each flight was flown in darkness, stars were the primary navigational means. Their importance to the Qantas crews is epitomised in the names given to each Catalina. On every flight, the navigator spent long hours at the rear fuselage blisters "shooting" stars with his sextant or using a drift sight to track flares. The flares were packed in metal tubes stored under the navigator's table and were dropped from the blister. On impact with the sea, they burst into smoke by day or flame at night, allowing the navigator to track the flare and determine wind speed and direction at that position using a drift/speed calculator.

Navigation was more difficult on Perth-Ceylon flights as the aircraft rarely flew above 2,000 feet because of headwinds, which meant that clouds frequently obscured the sun and stars. Eastbound, above 10,000 feet, many of the lower clouds were eliminated, allowing more frequent sextant observations. But the navigator could still spend long periods in the blister compartment, waiting for a gap in clouds to make a sextant observation of a star or the moon.

At the outset of the Indian Ocean service, a shortage of experienced navigators meant that RAF Ceylon agreed to loan navigators, with the Qantas captains filling in during rest periods. But the demands of the operation necessitated Qantas acquiring its own dedicated navigators. QEA Western Operations Division Manager Captain Crowther turned to the RAAF, requesting the release of navigators. Although he was anticipating experienced men, RAAF Western Area Headquarters at Perth promptly seconded to Qantas six recently trained navigators: Joe Bartsch, "Banjo" Patterson, Frank Sander, George Hoare, Dolph Nuske and Stan Pearn.

RAAF Western Area Navigation Officer Flight Lieutenant JLB Cowan, formerly a Catalina navigator with No. 20 Squadron in New Guinea, was concerned by their limited experience, none of which had been over the sea. Cowan invited a group of Qantas pilots from Nedlands to a meeting to discuss the situation, following which Crowther lodged a formal request for Cowan to be released as well, to supervise the navigation side of the operation. This was a pivotal move, as Jim Cowan was seconded to Qantas to instruct in navigation techniques in the classroom and during scheduled services. All six of the loaned RAAF navigators became skilled at the ocean crossings and some stayed on for long careers with the airline.

The only positive position fix close to the route was the Cocos Islands. At first the Catalinas avoided the Cocos by 40 miles because of military intelligence reports that Japanese bombers and flying boats made reconnaissance flights over the island group every few days. By late 1944 as the war situation improved and the RAF was constructing an airfield on West Island, the islands were overflown as a reassuring positive navigational fix almost exactly midway on the ocean crossing. On their normal schedules, the Catalinas passed over the Cocos Islands in each direction by night.

Now that the Western Operations Division had proved the Indian Ocean Service to be reliable, Qantas management more readily approved financial expenditure for the division. Captain Crowther was performing all the administrative work as well as being in operational command. He was now able to advertise for an administration officer to take over the bulk of this work, with Hudson Shaw being engaged. Captain Crowther told the author he had no recollection of crews ever complaining about the long hours and overloads. It was war, and all his men were playing their part.

Meteorology

A significant issue for the Qantas Perth-Ceylon service was accurate weather forecasting along the route and at the destination, over a day away. For the early flights forecasts were limited to the ports at each end because so little weather data across the ocean was available – there had been no previous need for such aviation forecasts.

Creating route forecasts involved investigating and predicting air circulations and winds in two hemispheres. The Cocos Islands were the only link in the route between Western Australia and Ceylon from which regular actual weather reports could be obtained. However, for a large part of the Indian Ocean there was scant information to assist forecasters. Qantas was fortunate to gain the services of two very skilled military meteorologists. Captain Ambrose reported:

> Wing Commander Grimes, who had been the Senior Aviation Meteorologist at Singapore before its fall, was now Senior RAF Meteorologist at Colombo. He came to Australia on the first RAF survey flight and conferred with the RAAF authorities and our company officials on the necessary weather organisation. The RAAF made available Squadron Leader Hogan, who established his headquarters in Perth. On his return to Colombo, Wing Commander Grimes appointed, at first, Squadron Leader Lea, and later Flight Lieutenant John, to Koggala to deal with our eastbound departures. There are few, if any, superior tropical forecasters to Grimes or Hogan, and we were fortunate in having their services.

G-AGIE being prepared for departure at Nedlands, showing the "Barker Bridge" extension to the jetty, which also carried the refuelling hose. (QEA)

A camouflaged Shell Oil refuelling truck at Nedlands in 1944. (Duncan Barker)

The Qantas Nedlands flying boat base in 1944, showing the tall maintenance hangar behind the jetty. The moored Catalina in the foreground is G-AGID. (Qantas)

A 1944 view of the Qantas Empire Airways flying boat base on the Swan River at Nedlands, Perth, showing the slipway eventually completed via a DCA works order. (Neil Follett Collection)

Broadly, Indian Ocean weather is frontal, governed by high-pressure systems which may be found in the southern part of the ocean, and which move across Australia in an easterly direction. Nearer India the northwest monsoon is dominant. It moves south with the Inter-Tropical Front, a belt of unsettled weather circling the world and moving south and north of the equator with the movements of the sun. The Inter-Tropical Front comprises low cloud and almost invariably rain, with occasional cumulonimbus cloud and lightning flashes. On many occasions, when flying through or near the Inter-Tropical Front, aircraft encountered the unusual meteorological phenomenon known to mariners from the earliest times as "St Elmo's Fire". Flying through such conditions could be alarming, particularly to unknowing passengers. The fire was a flaming brush-like electrical discharge, which occurred as the aircraft was flying among thunderclouds; it lit up engines and wingtips with an eerie blue light but caused no damage.

Meteorologists divided the entire Indian Ocean route into twelve zones for forecasting. The weather to be expected in each zone was detailed, as well as predicted wind velocities - a vital factor - at various altitudes. As each flight reached the halfway mark any updated weather details were broadcast to the aircraft. If, as was usual, this confirmed the original forecasts the captain could continue; it also gave him sufficient data upon which to make an abort decision. Records show, however, that very few services turned back because of an enroute change to the weather forecast.

The Cocos Islands, as a weather reporting station, contributed enormously to the predictions of the forecasters. In fact, Qantas made a special Catalina trip there in August 1943 shortly after their service began, to deliver two newly trained weather observers, who were there expressly for the Qantas service. This is described later in the Cocos Islands section.

Qantas crews carefully recorded actual winds and weather conditions encountered during each ocean crossing and passed that data to RAF meteorologists at Koggala and RAAF meteorologists at Pearce. The data collected enabled increasingly accurate forecasts to be prepared for each flight.

Captain Ambrose summed up the feelings of the division when he reported:

> In the opinion of the pilots of this Division, the meteorologists have done a really first-class job.

Airgraph Mail

At the beginning, the Catalinas were carrying 400 to 500 pounds of mail and diplomatic dispatches with each service. Much of the general mail was "troop mail" and "POW mail", being letters from families to Australian military personnel based overseas or prisoners of war. The "Airgraph Mail" was a wartime temporary measure to reduce the weight and size of mail. Instead of bulky mail bags of paper letters in envelopes, the Airgraph Mail system required letters to be written on special forms, which were then automatically numbered and photographed on microfilm rolls. A roll of film about four inches diameter could contain 20,000 letters. At their destination, the microfilm was printed out in letter format and sent to the addressees by normal postal services.

The concept had been suggested in early 1943 and was strongly supported at the British end by the Australian High Commissioner to the United Kingdom, SM Bruce. The first Airgraph Mail from England to Australia left on 4 June 1943, flown via the USA and reaching Melbourne on 22 June on board a USAAF Consolidated C-87 Liberator transport across the Pacific Ocean. Airgraph microfilms were processed by Kodak in Melbourne as printed paper letters, which were distributed by the standard Australian postal network.

It took some weeks to get the new mail system working smoothly. At first the rolls of film were sometimes not recognised as mail and misdirected or lost completely while enroute in Canada or the United States. But once settled in, Airgraph Mail proved a highly effective solution to the weight problem. As a morale booster, for the Christmas 1943 period, Qantas produced Airgraph Mail forms with artwork of a smiling jumping kangaroo carrying an air mail sack with the inscription "Qantas Empire Airways send you warmest Christmas Greetings from the land of the Kangaroo".

Airgraph Mail operated for fourteen months. It was discontinued from August 1944 because of the increased payload available with the introduction of the Qantas Liberators across the Indian Ocean. A QEA memo headed "First Through British Airmail England-Australia" states that the first general mail following the war interruption was postmarked in London on 24 August 1944. It arrived in Perth on 2 September on Catalina Service 1Q82 and was postmarked in Sydney two days later. The first outbound mail from Sydney was postmarked on 11 September 1944. It left Perth for Colombo on 15 September on Liberator Service 4Q10 and reached London on 25 September.

Nedlands

The Swan River at Perth proved a perfect base for the Qantas Catalinas. Its wide expanses of usually smooth water gave sufficient take-off distances for the heavily overloaded aircraft, the usual run commencing at the narrows near the city with a take-off run towards Fremantle.

The essential secrecy of the Qantas service was aided by the fact that Perth citizens were accustomed to the sights and sound of Catalinas on their river. The US Navy base at Crawley Bay had used the same stretch of river for over a year. The American PBY-5 Catalinas were matt black or camouflaged – the Qantas Catalinas in their RAF camouflage would not have stood out.

Senior Station Engineers Norm Roberts and Colin Sigley had an initial team of six engineers. This increased when the third Catalina was due to be delivered. A workshop was urgently needed, and an empty kiosk near the base site was rented and fitted out with three workbenches, shelving for spares stocks and a desk with a telephone. Roberts later wrote:

> It was indeed a great day in mid-October 1943 when we moved into the shelter of a building for the first time. We were now in command, for the first time, of our growing work commitment.

At first there were no refuelling facilities, so the Qantas Catalinas had to be taxied on the river around the headland two miles to the US Navy base

An Airgraph folded letter form with Qantas Christmas greetings for 1943. (Civil Aviation Historical Society)

at Crawley Bay to be refuelled. The secrecy under which the Qantas operation commenced extended to these visits to Crawley to refuel. A QEA crew member recalled the Americans carrying out the refuelling asking what they did to use so much fuel each time – the Australians' stock response that they were selling it on the black market seemed to satisfy! The assistance Qantas received from the US Navy at Crawley deserves special mention, Captain Crowther said:

> The US Navy helped us enormously; we just could not have got started without them.

The US Navy flying boat base at Crawley Bay had been established during 1942 for Patrol Wing 10, which had evacuated south after the fall of the Philippines. Its first Catalinas arrived at Perth in March 1942 and a sheltered scrub area on the banks of the Swan River at Crawley Bay was selected as the site for a base. Facilities such as sealed slipway and aircraft parking areas were built, and large open-sided maintenance hangars erected. The US Navy Catalinas provided protection to Allied shipping in the Indian Ocean as well as cover for US submarines

An aerial view of Crawley Bay in 1944. The two hangars of the US Navy seaplane base are at top centre while the Qantas Nedlands base is just visible on the other side of the peninsula, top right. (Qantas)

operating off the West Australian coastline. Patrol Wing 10 was redesignated as Fleet Air Wing 10 on 1 November 1942 and grew to be a large unit. One of the original squadrons, VP-52, which had been at Crawley since the first days of the Perth base, was still there when the base was closed in July 1944 as US forces moved north towards Japan.

Roberts gives the following account of the first encounters with the Americans:

> The existence of an American Catalina Base nearby on the Swan River was entirely coincidental but was fortunate for us - even though we had been warned not to seek help from them. Our first Catalina G-AGFM had been delivered to us in very good condition. It had been well cared for by BOAC and efficiently serviced by the RAF. Most of the components, including the engines, were original with adequate working lives still available for our immediate requirements. Our early servicing was therefore directed to functional testing and close examination of the aircraft in general - including of course changing the spark plugs. We did not have a spark plug spanner. This was a special tool and so it befell me to ignore the ban and approach the Americans for help. I borrowed a spanner on the promise it would be quickly returned. By this means we completed, while afloat, our first servicing routine on our first Catalina. We continued to borrow the spanner until one day I decided to show my gratitude more realistically to the crew chief with whom I had been dealing.

Before making my usual request, on this occasion, with a complimentary remark, I presented him with a couple of ebony elephants I had brought back from Ceylon. He looked at these quietly for a moment or two, and then said, with the first smile I had seen on his face "Man, you have just won yourself a spark plug spanner". Further exchanges occurred as the need arose, and the dreaded possibility of having to suspend our operations until a vital replacement spanner could be obtained was avoided during our developing period. It might be said that our flying at that time was often sustained by the power of an ebony elephant.

There is no doubt our confidence was further

bolstered as each successful operation was followed by another. So it was one morning that Bill Crowther and I were out in mid-stream standing together in our small boat. As we watched our departing Catalina heading out into the nor-western sky we impulsively turned and warmly shook hands. It was a spontaneous reaction to the sense of achievement that came from the despatch of our tenth service and that moment we felt our success was assured. The small boat we were aboard I later christened "Twinkle Star".

Roberts subsequently had the name *Twinkle Star* painted on the small dinghy with a six-pointed star just like the aircraft star names.

December 1943 saw the first engine change at Nedlands. As there was still no action from the DCA to construct a slipway, the Catalina was pulled ashore on loaned beaching gear at the US Navy base on condition that the exercise would be brief, to avoid any congestion affecting the Americans' operational activities.

The engine was quickly removed and transported to the Nedlands kiosk where it was placed on an engine-stand, which had been constructed there earlier. The Catalina was then towed two miles back to the mooring at Nedlands. The replacement engine was one of the new Australian licence-built Pratt & Whitney Twin Wasps. The power plant removed was reassembled in the kiosk. All hose lines were oil pressure tested before being refitted, and the electrics, such as generator and starter-motor, were overhauled, then bench-tested in the workshop of a motor garage in Hay Street in the city. The spares needed, such as new commutator brushes and bearings, were obtained from the Americans at Crawley Bay. Reinstallation of the power plant required beaching again at the US Navy base.

The Shell Oil Company installed aircraft refuelling facilities at Nedlands. The Shell aviation representative for Western Australia, Dudley Barker, devised a fuel delivery system for the Catalinas. This included a jetty to carry the fuel pipeline on to an end platform in deep water. A floating extension ramp for the jetty then carried a hose line to the aircraft and also allowed passengers and crew to walk aboard the Catalinas in comfort. The jetty rightly became known as the "Barker Bridge".

The first building on the Nedlands site was erected late in 1943. This long shed was the new workshop and store, replacing the original kiosk. It also was fitted with an office for Norm Roberts and a secretary. Roberts then drew up a proposal and design for a nose-hangar for Catalina engine maintenance and submitted this to QEA headquarters in Sydney. Chief engineer Arthur Baird supported the proposal and finance was approved. The completion of this hangar in December 1943 allowed the workshop activities to be moved out of the shed, which was then converted in part to a reception lounge for passengers.

The Nedlands base still did not have a slipway, although one beaching chassis set had been received from the RAAF. Slipway construction was the responsibility of the DCA, which was slow to respond to repeated Qantas requests. Norm Roberts wrote later:

> Our concern finally eased one day when several truckloads of blue metal and a work team arrived at the Nedlands base. Work on the slipway commenced and by the end of the day sufficient metal had been deposited and hand spread to form a nicely graded slope. It now only needed rolling and finishing with a concrete surface and we would be independently established.
>
> After vainly waiting a couple of weeks for this further work to proceed I impatiently rang the Works Department only to be told the job was complete. At my request, however, I was promised that an engineer would investigate the problem. He agreed that the ramp of blue metal would be useless for our purposes and explained that, unfortunately, it was all the DCA works order had specified. He did promise to urgently pursue the matter further.
>
> In desperation I did make a futile attempt to pull a Catalina up the ramp but almost immediately the wheels of the beaching dolly sank into the

blue metal, and I had to resort to using its engines to get the aircraft back into the water. The work on the slipway was completed some four to five weeks later, but in the meantime, I had temporarily surfaced the ramp with Marsden matting "borrowed" from the American base. It was by this means we were able to complete "on shore" inspections of our Catalinas at Nedlands. The matting remained in our possession and subsequently it enabled us to position aircraft beyond the limits of the tarmac for special work. It thus served on two occasions when several sheets of badly corroded bottom plating had to be replaced on G-AGFL and G-AGFM.

Koggala Lake

The RAF Koggala Lake station had been developed into a large seaplane base for maintenance and overhaul of Sunderlands and Catalinas. A runway for fighters was constructed on the edge of the lake. The RAF provided Qantas with an office for company traffic staff, as well as a stores shed for spares. Two station wagon vehicles were available for ground transport, with these being later supplemented by a staff car and heavy transport vehicles.

Initially BOAC had appointed one of their traffic officers to act as Qantas' station superintendent. His headquarters were at Koggala, and a booking office was maintained in Colombo. A few months later BOAC urgently needed his services elsewhere, so QEA's Perth traffic superintendent, RG Cochrane, was transferred to Koggala as his replacement. In turn, Cochrane's position at Perth was filled by AJ Quirk sent from QEA Sydney. Cochrane was later assisted in Ceylon by another traffic officer, John Day. Occasionally the Koggala staff made a road trip to RAF Trincomalee, the main RAF Catalina base on Ceylon, to collect spare parts from RAF stores.

Qantas flight crews at first stayed in the town of Galle, twenty miles from the lake, at the New Oriental Hotel, which had been taken over by the RAF as an officers' mess. Engineering staff stayed in the officers' quarters at RAF Koggala. The hotel proved noisy, and the situation was greatly improved from January 1944 when Captain Crowther arranged for Qantas to lease two houses at Closenberg, three miles south of Galle, to serve as a company crew accommodation. The manager of the New Oriental Hotel was contracted to provide hotel services, a dining room and lounge. The new Closenberg Mess was on a bay with a surf beach, which was greatly enjoyed by the resting Australian crews. It was also close to the RAF flying boat officers' mess in the old P & O line headquarters building.

The formality of the era is reflected by the Closenberg Mess Dress Code notice, dated 20 January 1944 and signed by Captain Crowther:

- For breakfast and lunch, shorts and shirts may be worn with shoes or sandals
- After 1800 hours, long trousers, bush shirts or tunic, shirt and tie must be worn
- The only exception to this rule is when officers arrive late from duty, in which case permission of the senior captain present must be asked to remain in working dress

Qantas personnel used other RAF Koggala facilities, such as the hospital situated at Galle. First Officer Rex Senior had a stay in the hospital in December 1943 for treatment of a severe tropical abscess. The secrecy of the Indian Ocean service was maintained, First Officer Ivan Peirce recalled:

> At Koggala at the beginning only the CO and Senior Ops people knew what we were doing. We used to drink in the same mess as the RAF pilots, but they didn't know we were flying to Australia.

Operationally, Koggala Lake proved to be a tight operation for the heavily laden Qantas Catalinas departing for Australia even though the RAF had built a seawall to raise the water level and hence increase the area of the lake. The Qantas take-off technique for calm wind conditions involved taxiing to the far south-eastern corner of the lake, and back-tracking up a creek to obtain the maximum possible take-off distance. These long take-off runs required the engines to be kept at maximum take-off power for well over two minutes, which was beyond normal engine operating limits.

Water-taxiing in the shallow waters was hazardous due to large rocks – an effective warning was a derelict RAF Catalina lying in mangroves alongside the creek!

G-AGID Rigel Star taxiing for departure at Lake Koggala, Ceylon. (Qantas)

Catalina G-AGFM passes the ancient town of Galle, close to Lake Koggala. Qantas flight crews initially stayed at the New Oriental Hotel, Galle, which had been taken over as the RAF Officers' Mess. (Imperial War Museum)

Take-offs from Koggala were so critical that to reduce drag, the wingtip floats were retracted at 40 knots. If the water was flat calm, the RAF crash launch would pace the accelerating aircraft on its take-off run, then cut across the bow at the right time, to break up surface water with its wake, to help the Catalina "un-stick". The frequent tropical rainstorms sometimes caused take-offs to be delayed, because the weight of the rain on the airframe affected take-off performance. It was not uncommon for a Catalina to abort several take-off runs at Koggala, before finally getting away for the long haul to Perth. Normally at least one hour was needed to allow time for the engines to cool, after an aborted take-off.

Karachi Extension

An unsatisfactory part of the operation quickly became evident: the slow air connection between Qantas' Ceylon terminus and BOAC's turn around point at Karachi. The connection from Colombo, via Madras and Bombay to Karachi often took two days or more.

QEA Managing Director Hudson Fysh recorded in his diary the existing cumbersome air connections between England and Ceylon during his homeward journey from his 1943 London visit. He departed England in August on board an RAF Transport Command aircraft, which flew well out to sea around

RAF Korangi Creek at Karachi was the terminus for the Qantas Catalina service from Australia. This 1945 view shows a busy scene of RAF Catalinas and Sunderlands. (RAF)

Spain before circling and landing at Gibraltar. It then continued to Cairo and Jerusalem with lengthy waits at each stop. At Jerusalem passengers and cargo were transferred by road to an RAF station on the Dead Sea, from where an RAF Catalina made the flight to Karachi. There Fysh boarded a Tata Airlines Stinson A trimotor to Bombay, with a stop at Ahmadabad, where the passengers waited while the pilot had a shower. The last leg from Bombay to Colombo was in a single-engined Waco biplane operated by another Indian airline.

A few months after the Indian Ocean service commenced, BOAC asked Qantas to extend the route to Karachi (then an Indian city), where passengers and mail connected with the scheduled BOAC main route to England. The delivery to Perth of the next two Catalinas G-AGID and G-AGIE during September 1943 enabled Qantas to agree to the Karachi route extension - which added 1,490 miles to the service. For safety reasons the Catalinas flew over the sea rather than by the more direct overland route.

The Karachi terminus was the RAF Korangi Creek flying boat base, situated a few miles south of the city. It was a former Imperial Airways/BOAC station on the edge of the Korangi Creek waterway, taken over by the RAF as a flying boat base. BOAC maintained a civilian terminal at the RAF station, operating Short Empires on the "Horseshoe Route" to Africa and Short Sunderland IIIs to England. The BOAC personnel provided all the handling functions for the Qantas Catalinas.

G-AGFM *Altair Star* made the first Karachi flight, Service 2Q18, during the night of 4-5 November 1943, commanded by Captain Crowther with First Officer JW Solly. The aircraft had arrived at Koggala from Perth on 28 October and was serviced prior to departure. Flying time was 12 hours 30 minutes at night, alighting at Korangi Creek, Karachi, at dawn. Here the passengers and all-important mail and diplomatic dispatches were transferred to the BOAC Sunderland service to England via Cairo.

The first east-bound flight from Karachi to Ceylon, Service 1Q18, was made by the same aircraft and crew, departing Karachi on 6 November and reaching Perth three days later, after a delay due to bad weather across the Indian Ocean. The total distance of 4,970 miles was covered in a flight time of 41 hours 40 minutes, with one stop enroute at Koggala Lake.

Catalina G-AGFM Altair Star at Karachi loading passengers from a small boat. (Imperial War Museum)

Qantas operations into Korangi Creek were scheduled for daylight hours because of the risk of collision with the many small boats always active at the mouth of the Indus River. The main harbour at Korangi Creek runs north south while the base water area was east west. As Captain Tapp recalls:

> When the wind was too strong for use of Korangi Creek we taxied into the main harbour and great care was required to avoid ships, sampans, etc.

The Perth-Karachi route settled into a routine. The aircraft were scheduled to arrive at Koggala in the early daylight hours from Perth. They were serviced during the day prior to the night flight to Karachi that same evening. The Catalina usually remained at Korangi Creek for 30 hours, during which it received a routine service by BOAC engineering staff. Parts were not a problem because the RAF station was well stocked for routine servicing of RAF Catalinas. Eastbound Catalinas departed Karachi at 1800 and landed at Koggala at 0700 the next day. Two hours later the flight continued with a new crew, alighting at Perth between 1000 and 1200 the following day. There the airmail and onward travelling passengers were driven to Maylands aerodrome (replaced from 1944 by the nearby newly built Guildford aerodrome) to connect with an Australian National Airways DC-2 or DC-3 passenger service to Adelaide, Melbourne and Sydney.

Qantas crews performed almost super-human feats to maintain the service, including the need to be almost constantly alert for the average 28-hour duty period necessary to accomplish each Indian Ocean crossing. At least one respected captain had to be taken off the service owing to his inability to stay awake during the lengthy flight. Crews had a roster, whereby they interchanged duties during the flight, on the premise no crewmember would have more than four hours of continuous duty. However, in a very noisy un-soundproofed Catalina, the two bunks were not conducive to restful sleep. Captain Ambrose, in a company report, detailed crew and aircraft procedures:

> With these increases in service came an increase in crews and the Karachi extension automatically doubled the time crews were

away from Perth. The original plan had been for each team to fly from Perth to Ceylon, rest there for several days, and then return to Perth. Now, each crew having brought a service from Perth to Ceylon, remained there as standby crew, whilst the previous standby crew that afternoon took the aircraft on to Karachi. Arriving there at dawn they left on the return flight in the late afternoon of the following day. This meant a dawn arrival at Koggala, and the next standby crew were at base ready to leave for Australia as soon as the load was changed to their aircraft.

By this means each crew was rested on the outward as well as the homeward voyage. Aircraft on leaving Perth went through to Karachi and returned to Koggala, where they were beached and serviced before returning to Perth one service later. It was intended that each crew should do one round trip per month, and as the average trip took about 80 hours, thus complete something over 900 hours per year. In fact, due to sickness and a variety of other causes, most crew members did much more time on the route, at the end of the first year, more in the region of 1,100 hours.

At Karachi, Qantas crew members initially stayed at two hotels, one of which was The North Western. Later they were billeted at the Karachi Museum, which was used as a mess by BOAC for passengers and crews. QEA navigator Joe Bartsch later wrote:

> Wartime Karachi was a very busy city, as it was the main western port in India and accommodation was exceedingly scarce. We stayed in the museum in the main hall, all the animal exhibits including an elephant we moved to one end, and we were sleeping in cubicles at the other end. But we were so pleased to have a good night's rest, even though there were tigers and an elephant in the room!

Things improved considerably when BOAC moved their Karachi mess from the museum to the former Japanese Embassy. Despite the irony of the situation, Qantas crews were pleased to move with them.

To maintain services on the extended Perth-Ceylon-Karachi route, Qantas was always looking for additional suitable aircrew. Among those joining the Western Operations Division during 1944 was Captain RO Mant, formerly flight superintendent of WR Carpenter Airlines' Sydney-Rabaul service flying DH.86Bs and then Lockheed 14 Super Electras until the Japanese invasion of New Guinea. Dick Mant, who would go on to an impressive career with Qantas, later recalled:

> I joined Qantas in May 1944. Captain Lester Brain, the manager in Brisbane offered me a choice of either captain on Lodestars to New Guinea on the American courier service, or first officer on Catalinas in Perth. I had done a lot of flying in New Guinea, so I preferred going to Perth on the Catalinas – it was something different and perhaps more challenging. Qantas sent several of us recruits to RAAF Rathmines and we lived there while we did a Catalina conversion course. The air force training was quite good, although I thought their night flying part was a little sketchy.
>
> I flew on the Catalinas operating from Perth to Karachi for 24 months, three trips as first officer, then captain again. It was a nonstop run to Ceylon in radio silence, and you had to be able to navigate sufficiently well to relieve the navigator when he had rest periods. The Catalinas were flown at a constant airspeed of 110 knots in order to achieve maximum range. As the fuel load burned off and the speed crept up, we kept reducing power to get the airspeed back to 110 knots.
>
> We would leave Perth at 10am and arrive in Ceylon at 7am the following day. Coming back, we would leave fairly early at 7am and touch down on the Swan River at Perth at 10am the following morning. The schedule took us over the Cocos Islands in the middle of the night. The Japanese used to fly over the Cocos and we didn't cherish running across enemy aircraft out there in daylight. Each pilot did about three return trips from Perth to Karachi each month. I think I enjoyed flying the Catalinas more than any other flying I did.

Passengers

The loading for each service had to be strictly controlled and was determined in the following manner. The Australian and Indian Departments of Civil Aviation arranged passenger priority while the armed services were granted a mail loading to an agreed limit, with government dispatches treated similarly. In many cases Qantas did not learn the names of their passengers until the day of travel. The airline was only advised of the number of passengers travelling on the next service. Not generally known is that the Indian Ocean service was one of the principal channels for the shipment of mail to Australian prisoners-of-war in the European and North African theatres - an average of 50 pounds of POW mail was carried on each Qantas service leaving Australia.

Wartime travel was severely curtailed for Australian civilians. To fly across the Indian Ocean as a passenger required government permission, which was generally restricted to essential travel in the national interest as part of the war effort. Intending passengers submitted their applications with supporting documentation. A Department of Civil Aviation high-level secretariat was advised of approved passengers with their requested travel dates. DCA in turn booked their passage with Qantas. Thus, essential secrecy was maintained. Once the airfare had been received, Qantas issued documentation, including a Qantas passenger ticket, shortly prior to travel. The ticket, a standard airline document, covered the normal airline conditions of carriage even though subject to wartime restrictions because it was still civilian airline travel.

The majority of passengers were government officials and senior military officers, including a number of American generals. From the crews' point of view, they were good passengers; they always pitched in and helped with tasks during the flight and often prepared their own meals and hot drinks, expecting none of the comforts of pre-war airline travel. British Lieutenant General Charles Gairdner, at that time Winston Churchill's personal emissary to Douglas MacArthur, Supreme Commander of the South West Pacific Area, used the Catalinas on his

The view looking backwards during take-off from G-AGFM. (Neil Follett Collection)

missions to Australia. Occasional celebrities carried included the popular British actor and singer Noel Coward on an entertainment tour for the troops.

Three passenger chairs, removed from Short Empire flying boats, were installed in the blister compartment. Two crew bunks in the next compartment forward were for aircrew rest breaks during the flight. Occasionally a passenger would occupy a bunk and usually this was quietly accepted. The toilet was attached to the rear bulkhead without any enclosure, regrettably giving little privacy.

Very few women passengers were carried on the Catalinas. The first was prominent British MP Dame Edith Summerskill who flew from Perth to Ceylon in July 1944 while returning to England after an Empire parliamentary delegation. Captain Crowther was concerned for her comfort on the long flight and, in particular, the very basic toilet facilities in the aircraft. He discreetly arranged for his lady secretary in Perth to meet Dame Edith

One of the few Catalina lady passengers, Dame Edith Summerskill and fellow British politician J Harris pose with Captain JAR Furze before their departure from Perth on 31 July 1944. (Qantas)

The Secret Order of the Double Sunrise certificate.

before departure and explain the situation. In any event she proved a popular passenger, saying she thoroughly enjoyed her flight in G-AGID *Rigel Star*, Service 2Q74, which departed Nedlands on 31 July under the command of Captain Furze.

For their long flight each passenger was issued with a flying suit, fleece-lined boots and blankets. Reading material was not provided because of weight considerations, passengers being told to bring their own. At night the aircraft was blacked out. Catering arrangements were salads and cold meats; cheese and biscuits; cake and chocolate (the latter being a wartime luxury, subject to food rationing) and hot soup. On early flights hot tea and soup was provided in large vacuum flasks, but as the service settled down, it was agreed the extra weight of a small electric hotplate was acceptable. The hotplate was fitted in the blister compartment to allow tea and coffee to be brewed. Usually, it was the lot of the second officer to serve in-flight refreshments. The hotplate allowed extra variety with cans of baked beans or spaghetti being warmed up and served. The food was carried on board in a company wicker basket, which had an allotted weight of 75 pounds. However, one captain recalled occasions when several ground staff were needed to lift the basket into the aircraft!

In recognition of a passengers' ocean crossing, Fysh devised a special memento of their flight. It was a coloured, gaily decorated certificate of membership in the *Secret Order of the Double Sunrise*. When disembarking at their destination, each passenger was presented with the paper certificate signed by the captain.

Safety

In any airline operation safety is paramount. The Qantas Indian Ocean service included a unique safety feature: prior to operating each Catalina or Liberator ocean crossing, the aircraft was test flown. The test flight was usually conducted the day before the scheduled service, with a full crew to ensure all positions were serviceable. With an ocean sector of over 3,000 miles, nothing could be left to chance.

The only alternate landing site in case of emergency, once clear of the Australian coastline, was the Cocos Islands. Christmas Island, well east of the route was a consideration but it was Japanese occupied. In the event of a forced alighting, individual life jackets and a nine-man life raft, equipped with food and water, emergency radio transmitter and first-aid supplies, were carried. Emergency drills were practised early on each flight. On every crossing, at dusk the aircraft was blacked out to reduce the chance of sighting by enemy aircraft.

The reliability of the Pratt & Whitney R-1830 Twin Wasp radial engines which powered the Catalinas is legendary. They were 14-cylinder twin row radial engines developing up to 1,250 horsepower. The same engines were also used by the Liberators, which were to replace the Catalinas across the Indian Ocean. Qantas experienced only six in-flight engine shut-downs in nearly 600 Indian Ocean crossings – a remarkable statistic.

Twin Wasp engines on the Nedlands hangar upper deck during a Catalina engine change. (Qantas)

BOAC supplied the Catalinas and later Liberators with original American-built 1,200 horsepower P&W R-1830-92s. During routine engine changes early in Qantas service, these were replaced by Australian licence-built R-1830-82 Twin Wasp engines producing the same power. The Australian production was by the Commonwealth Aircraft Corporation at a Department of Aircraft Production factory in the Sydney suburb of Lidcombe. A total of 870 Twin Wasps were built by CAC between 1941 to 1945, before the Lidcombe works converted to Rolls-Royce Merlin production. Earlier CAC had built 680 single-row Wasps at Fishermans Bend, Victoria, between 1939 to 1943.

G-AGID *Rigel Star* was the first Catalina to be fitted with two Australian-built engines during February 1944. When the engines required major overhauls, they were removed from the Catalina in the Nedlands hangar and packed for rail transportation 300 miles to Kalgoorlie. Here the work was carried out at RAAF Station Boulder in the engine repair workshops of No. 4 Aircraft Depot. Roberts later stated:

> To the very worthy credit of the RAAF overhaul depot at Boulder we enjoyed the utmost confidence in the dependability of our engines over the Indian Ocean throughout the full period of our Western Operations.

The US Navy at Crawley Bay also sent the engines of their PBY-5 Catalinas by rail to RAAF Boulder. No. 4 Air Depot was a busy maintenance unit with almost 1,000 personnel by May 1944 when aircraft test flights were averaging 24 per month in addition to an average of six ferry flights. The depot report for 27 May listed its workload for that day as eighteen aircraft (major inspections, modifications or crash repair), 22 Pratt & Whitney R-1830 engines, 28 magnetos, eleven carburettors, 28 hydraulic pumps and six propellers.

Cocos Islands

The Cocos (Keeling) Islands, usually referred to as the "Cocos Islands" or just "Cocos", is a small Indian Ocean island group comprising 27 coral islands. The Cocos Islands played a significant part in the Qantas Indian Ocean operation: initially sending invaluable mid-ocean weather reports and later, after the construction of an RAF airfield in 1945, as a staging point.

The islands are a typical coral atoll, horseshoe shaped, consisting of five main islands: Horsburgh, Direction, Home, South and West Island. They are from one to five miles long with a ridge of twenty feet above sea level and are surrounded by an almost complete circle of coral. The islands enclose a lagoon seven miles wide and nine miles long. Their most important asset, however, was their strategic location approximately halfway between Perth and Ceylon. The nearest land is Christmas Island some 350 miles east, which was occupied by the Japanese. Neither they nor the Allies attempted to set up an operational base on Cocos Island in the early stages of the Second World War - both sides were unwilling to divert forces to defend such a remote spot. But Japanese reconnaissance aircraft regularly overflew the Cocos on their patrols.

The sociological history of the islands is complex and fascinating. They were discovered by Captain Keeling of the East India Company during 1609. The first permanent settlement occurred in 1827 when John Clunies-Ross arrived with his wife and family, setting up a family dynasty which brought in Malayan plantation workers for the islands' coconut trees. The Cocos Islands was declared a British colony in 1857 with John Clunies-Ross as governor. They were later annexed to Ceylon before being transferred in 1886 to the administration of the Straits Settlements at Singapore. British administration from Singapore was to end in 1955 when the Cocos (Keeling) Islands were transferred to Australia, finally becoming Australian Territory in 1972.

A cable station was established on Direction Island in 1901 during the laying of the undersea cable to allow telegraph communications between Australia, Singapore and Africa. In 1910 a Wireless Station was added to communicate with passing ships and send weather reports. Early in the First World War, in 1914 a landing party from the German cruiser *Emden* came ashore to destroy this installation. A radio message was sent from Cocos, and the cruiser HMAS *Sydney* was rushed to defend the islands. After a furious engagement the German vessel was beached on North Keeling Island: HMAS *Sydney* had secured Australia's first naval victory.

The Cocos Islands were affected by war for a second time in 1942 when a Japanese cruiser shelled the cable station and departed in the belief the installation was destroyed. But the attack in fact caused relatively minor damage to the station. A successful Allied bluff convinced the Japanese that Cocos was no longer worthy of attention. This allowed regular weather observations to be sent via the cable to Broome, then Perth. These were vital for accurate route weather forecasting for the Qantas operation.

Captain Lewis Ambrose (arms folded) and the crew of Catalina G-AGID pose with army men after the first Qantas landing at the Cocos Islands on 30 August 1943. The photo was received via First Officer Rex Senior who is on the right-hand side holding coconuts.

During the period of the Qantas Perth-Ceylon service, their Catalinas alighted at Cocos on only four occasions, each time in radio silence. Under wartime security, Cocos was assigned the code-name "James" and flights to Cocos are recorded in the QEA aircraft logbooks under this name. The first Qantas visit occurred shortly after the Indian Ocean Service commenced. During the night of 29-30 August 1943, Captain Lewis Ambrose made a special flight from Ceylon to Cocos in G-AGID *Rigel Star*. The purpose of the flight was to drop off two Perth schoolteachers who had been trained as meteorological observers and were to be stationed on the islands for the duration of the war. As well as teaching at the Cocos school, they would send regular weather reports to Perth via the cable.

The story of these two meteorological observers highlights the difficulty in locating the Cocos Islands by the aerial navigation of the time. The original plan was to fly them from Perth to Cocos on board a Catalina from US Navy Patrol Wing 10. A fully armed PBY-5 equipped with radar was flown to the US Navy base at Crawley Bay from Exmouth Gulf. The first sortie to Cocos was unsuccessful - the US Navy crew could not locate the island group and the aircraft returned to Perth. Several days later a second attempt was made and again they were unable to locate Cocos. After searching for some time, the US Navy crew called their base at Exmouth for a radio fix, but because this transmission might have been picked up by the enemy, they left the area and made for Exmouth. Approaching the West Australian coastline with fuel exhaustion imminent, equipment was jettisoned, including firearms and their hapless passengers' luggage. The Catalina finally made a forced landing in heavy seas several miles short of Exmouth and was towed into Exmouth Gulf by boat.

The delivery of the meteorological observers was now handed over to Qantas with Captain Ambrose placed in charge of the operation. He decided there would be a much better chance of locating Cocos by flying south from Ceylon, so the two men were flown from Perth to Koggala Lake as passengers on a scheduled Qantas service. Not wanting to disrupt the scheduled services, Ambrose took the opportunity of the pending delivery of their next Catalina G-AGID from England to Ceylon, which reached Koggala Lake on 26 August. Senior Navigator Jim Cowan was to have gone on the Cocos flight, but Ambrose decided against this because he wanted to save the extra weight for fuel; he preferred to carry out his own navigation so that in the event of failure to find Cocos the blame was his alone. Cowan was instead assigned as navigator on the next scheduled Ceylon-Perth service.

G-AGID departed Koggala as planned on 29 August 1943 at 1800 with a minimum crew under Captain Ambrose and First Officer Senior, carrying the two travel-weary meteorological observers. If Cocos could not be located after an hour's square search, they would have insufficient fuel to return to Ceylon, so a diversion was planned to the Diego Garcia island group to the west. All went well - at dawn Ambrose made a run into Cocos along a navigation "sunline" and in his own words:

> We hit Cocos right on the nose. It was good to see North Keeling Island close on the port hand and know we had succeeded.

After refuelling, Ambrose departed Cocos for Perth, logging a total flight time of only 28 hours for the journey Koggala-Cocos-Perth, a very short time given that it included the two climb-outs.

Catalina Bombed

The only occasion a Qantas aircraft on the Indian Ocean service actually encountered the Japanese was at Cocos, when a Catalina was bombed. Captain Russell Tapp had been tasked to pick up a naval officer who had been put ashore by a warship earlier who was suffering acute appendicitis and needed surgery.

Flying G-AGFL *Vega Star*, Tapp departed Perth on 14 February 1944 at 1746 with a crew comprising First Officer Senior and four others. After a flight of 14 hours 26 minutes, they alighted at 0812 the next morning and moored near the jetty at Direction Island. Captain Tapp described the events that followed in his company report:

> The Cocos Islands are a very small spot in the middle of a very big ocean. Because failure to locate the islands would also mean failure to refuel, our aircraft left Perth with full fuel tanks and the decision made that when, according to our navigation, we should be over the Cocos Islands, we would spend one hour trying to locate them. If, after that time, we still had not located them we would continue to Ceylon. One hour was the maximum time we could remain over the area and still have the minimum reserve of fuel for the arrival at Ceylon.
>
> The flight was planned as to arrive at Cocos 90 minutes after sunrise and for the approach to be made along a "sunline". This means that a track is to be flown which will place the sun exactly at right angles to the track so that position lines can be obtained running directly along the track of the aircraft and, if navigation has been exactly correct, should pass through the islands. Two positions which satisfied these conditions existed, one to the north of the island and one to the south. The flight being made from Perth the shortest route was to approach from the south so that a course was planned well to the south of Cocos.
>
> Altitudes of heavenly bodies read with a sextant are unreliable until the sun has risen well above the horizon, hence the planned arrival at Cocos 90 minutes after sunrise. This provided for satisfactory readings to be taken for the half hour before arrival - approximately the last 50

miles of the run in. The flight went according to plan. We arrived at the position, by our navigation, where we should alter course for Cocos and did so. After one hour's flight we "shot" the sun at three-minute intervals, plotting each position line. As we were flying low and below some broken cloud, conditions were not quite smooth and minor variations resulted in the position lines obtained, the average however indicated that we were 10 miles to the east of the desired track. The course was therefore altered 90 degrees to port for 5 minutes then back again to the original course. Two minutes after our planned arrival time at Cocos we were over the islands and landed on a lagoon.

So as to arrive at a suitable time in Ceylon, departure was planned for dusk that evening. As soon as the cable station staff came on board, they asked us for our plans and said that we were quite safe as the regular Jap reconnaissance aircraft had been over two days before and that these flights were made approximately every ten days. We were, therefore, quite safe for the day. Breakfast was ready ashore. The day was ours - we could swim or pay a visit to the other islands in the group or fish - what would we like to do? I said first we would have breakfast, then we would refuel the aircraft ready for departure, after which I proposed to have a sleep until lunch time and then decide what to do for the afternoon. Everyone agreed with this proposal, and we went ashore where an excellent breakfast was waiting for us.

Breakfast finished, the remainder of the crew went back on board to carry out the refuelling. I followed some 20 minutes later to see that everything was in order. The manager of the cable station had told me that the locals were a bit jittery about aircraft. Direction Island is only a small island and the noise of the sea-breakers on the other side of the island was often mistaken for that of an aircraft engine and the locals often became very excited thinking that an air raid was about to happen. As we strolled from the Cable Station towards the jetty, we heard a noise which resembled an aircraft engine. Some commotion was immediately seen amongst the locals, and a bell was rung. Barker, the manager, had just said "You see what I mean", when the noise quickly increased and there was no doubt that it was an aircraft.

The Jap came around the corner of the island at about 1,500 feet altitude - it appeared to be what was known as a "Betty" bomber - and flew straight over the lagoon where the Catalina was anchored. I think the Jap pilot was as scared as everyone else because, instead of dropping any bombs or firing guns, he suddenly altered course a few degrees, opened the throttle full out and climbed away as quickly as possible. It was obvious that he was not prepared for any hostile aircraft in the islands and was not ready to release bombs. He disappeared over the other side of the island, made a circuit and came back at about 3,000 feet, flew overhead again and released two bombs. We, of course, lay flat on our faces. Most of the crew of the aircraft dived overboard. Both bombs, although making plenty of "swish" sound while falling, missed the aircraft, the nearest being some 70 feet away, and fell harmlessly into the lagoon. No shots were fired, and he disappeared again over the island. but we could still hear his engines.

Everybody from the aircraft came ashore and we waited for the next attack, but it did not come. After a while we could no longer hear the engine and hoped he had gone away, probably, we thought, to fetch reinforcements from Christmas Island, if there were Japanese aircraft stationed there (which nobody knew), or perhaps he would signal Java to send out aircraft from there.

It was inadvisable and impractical to leave the island at that time so decision was made to off-load everything on board so that if the Catalina was destroyed everything removable would be safe. This done we could do nothing but wait. It transpired that the naval specialists had installed a radar screen on the island, and this was the first time it had been used. He could still see the aircraft on the screen but about half an hour later he had lost it, so we presumed he had left the area. Everything was left as it was until the afternoon. No other aircraft having arrived, we refuelled our aircraft, placed the load back on board and left at dusk, arriving at Ceylon as planned.

With the new passenger aboard, Tapp took off at 1815 and alighted at Koggala Lake at 0727 the next morning, after a flight time to Ceylon of 13 hours 12 minutes. Those on board, including the unfortunate naval officer, considered themselves fortunate the Japanese aircraft had not pressed the attack more vigorously. Although two-thirds of the Indian Ocean route was operated through Japanese patrolled territory, this was the only time an enemy aircraft was sighted by a Qantas crew.

Night Alighting at Cocos

A final unplanned visit to Cocos came a year later, when Captain OFY Thomas flying G-AGID *Rigel Star* from Perth to Ceylon was forced by engine trouble to divert and make a night landing on the lagoon. That morning, 28 January 1945, the crew of six had assembled at the QEA Perth city office at 0700:

>   Captain OFY (Frank) Thomas
>   First Officer HG (Harry) Mills
>   Second Officer AL (Ash) Gay
>   Navigation Officer Howard (Joe) Bartsch
>   Radio Officer Warren (Nobby) Clarke
>   Flight Engineer FJ (John) Maskiell

The "met" briefing from Squadron Leader John Hogan forecast tail winds as far as Cocos. The crew was driven to Nedlands and with no passengers, *Rigel Star* departed the Swan River at 0815. The winds were as predicted and late that afternoon, running ahead of normal timings they had a rare view of the Cocos Islands in fading light as they passed 10 miles clear. Normally the Catalinas passed the Cocos during the night in each direction. Captain Thomas had been concerned by instrument readings for the starboard engine and had switched magnetos off in turn. One hour and 20 minutes past Cocos, the starboard engine suddenly lost oil pressure. Thomas shut it down and feathered the propeller. He decided the twelve more hours to reach Ceylon was too great a risk on one engine and instructed navigator Joe Bartsch to calculate a course to get them back to Cocos. Bartsch takes up the story:

> Fortunately, the weather was fine and clear, so I was able to get a suitable three star fix to plot a course and was relieved when Ash Gay reported that he could see the moon shining on the island's lagoon. As we had radio silence and the island was in black-out conditions, we decided to use our flame floats to light the landing run. Harry Mills and I, sitting on each side of the blister, dropped them on the lagoon at 2 second intervals. When the captain turned in, we had laid an excellent flare path. He then did a perfect alighting and taxied towards the wharf at Direction Island. We were met by a launch with four armed soldiers, but they soon recognised the Qantas aircraft and helped us mooring the Catalina.

> As our flare path burnt for 40 minutes, it was fortunate there were no Japanese aircraft in the vicinity at that time.

The islands were under British military control and Colonel Jessamine was the officer in charge of a detachment of soldiers. While he took Captain Thomas to the cable station on Direction Island to advise Perth base about our diversion, the rest of the crew were taken to Home Island for a great meal and clean quarters in wartime Nissen huts.

On Monday 29 January we all went back to the Catalina to hand refuel it from a stack of four-gallon drums. Meanwhile the flight engineer had found a loose fitting on the oil line that had caused the engine trouble – he soon had it fixed and refilled with oil. Captain Thomas decided to give the aircraft a taxi check, and all went well on the lagoon, but when returning to the wharf the engine spark plugs oiled up. So the engineer had to change the eighteen plugs in each engine;

one of the few spare parts we carried with us. To do this he had to erect a collapsible platform and hang it on the leading edge of the wing so that the cowlings could be removed. During the exercise with many plugs out, he dropped the plug spanner into the water. So now we were really marooned!

We had a swim in the lovely clear lagoon but were unable to find the spanner. The captain sent another cable to Perth, this time requesting more plugs and a replacement spanner. On Thursday 1 February we were disappointed to learn our supply plane from Perth had been delayed 24 hours by weather. Fortunately, it arrived early the next day with our supplies, also items requested by Colonel Jessamine.

The relief Catalina was G-AGIE *Antares Star* flown by Captain Bert Ritchie, who took fourteen hours from Perth. He moored alongside *Rigel Star*, bringing the requested engine parts and two Nedlands ground engineers, A Hall and W Stitt, to assist with the repairs. Although the ruling Clunies-Ross family were in England for the duration of the war, as a gift for their hospitality, Ritchie brought a collection of fishing lines and nets, which were in short supply on the islands. To resume the schedule as quickly as possible, Captain Thomas' crew took over the relief Catalina to carry on the service to Ceylon. Joe Bartsch continues:

> We taxied to the southern part of the lagoon and commenced our take-off run towards Direction Island. Just as we were about to lift off, we hit a submerged coral head that punched a hole in the compartment between the radio officer's chair and my navigator's chair and a jet of water poured in between us. The situation called for a quick decision by Captain Thomas, who decided to continue the take-off. The flight to Koggala on the southern tip of Ceylon continued without further trouble and when nearly at our destination, we radioed the need to have the Catalina beached as soon as possible after alighting. Nobby Clark and I jammed a flat toolbox and several flight manuals against the sprung plate, forcing it back into position. There were two motor launches to meet us on arrival, and we were soon towed ashore with beaching gear bolted to the fuselage.

Captain Frank Thomas had exhibited airmanship of the highest calibre. Ritchie took G-AGID back to Perth for a full check-over before her next operational crossing. The British military detachment was making preparations for a military airfield to be built on Cocos. Two months later a shipping convoy arrived from India carrying airfield construction equipment and a force of British and Indian army, navy and air force personnel to build an RAF airfield on West Island.

A Geraldton Diversion

On 18 May 1944 Catalina G-AGFL *Vega Star* operating eastbound service 1Q59 diverted to Geraldton, a coastal town 200 miles north of Perth. As dawn broke after the long night flight from Ceylon, smoke could be seen billowing from one engine and Captain Don MacMaster shut it down and feathered the propeller. After flying another 800 miles on one engine he decided on a precautionary landing at Geraldton. The shipping harbour at Geraldton had rudimentary facilities for flying boats because it was used as a refuelling point by US Navy single engine floatplanes flying between their bases at Exmouth Gulf and Crawley Bay. The Catalina was carrying two military passengers, Lieutenant Colonel Lyon and Major Tims, plus a large load of mail. MacMaster alighted off Geraldton harbour after 28 hours 27 minutes flying time from Ceylon, slowed by the engine failure. Passengers and mail were transferred by boat to shore, to continue to Perth by road.

Flight Engineer Wally Martz, who incidentally held the record for making the greatest number of crossings on the Qantas Indian Ocean Service, inspected the offending engine. He first checked a cylinder in the R-1830 rear row, which had previously been leaking oil – and could see that it was now leaking heavily. Martz tightened the base hold-down bolts at the base of the cylinder and following crew rest in Geraldton, they ferried G-AGFL empty to Nedlands. Whisps of smoke from the oil leak had been earlier noticed several days earlier by Koggala Lake traffic superintendent RG Cochrane on that aircraft's arrival from Perth. The base nuts were tightened by the Koggala station engineer and the engine tested. The Catalina then made a return trip to Karachi without any further trouble. On this basis, Captain MacMaster accepted the aircraft for the service to Perth - it was only after more than 24 hours of continuous engine running, that the problem manifested itself again.

At Nedlands the offending engine was removed and sent for overhaul to the RAAF at Kalgoorlie. On instructions from Norm Roberts, it was to be stripped to components and set aside until he could visit to inspect the internal parts. Roberts found the cylinder had a bent master rod. Records revealed that the master rod was a replacement unit fitted during the previous RAAF Boulder overhaul. Checking further, Roberts found that the rod was from reconditioned stock, rather than new unused - as stipulated by the Qantas contract.

Very few Catalina services were aborted, but it did happen occasionally. Service 2Q35 departing Perth on 1 February 1944 was operated by G-AGIE *Antares Star* under the command of Captain RJ Ritchie. Carrying three passengers plus mail, he was forced to return to Perth owing to the port engine losing power. When they alighted back at Nedlands the aircraft had been airborne for 17 hours and 33 minutes. After repairs, the same aircraft and crew resumed 2Q35 four days later. Captain Bert Ritchie went on to a long Qantas career and was appointed general manager in 1967.

In April 1944 Western Operations Division Manager Captain Crowther sent a memorandum to all division staff, highlighting their achievements. Stifled by wartime censorship, it read:

> I have pleasure in quoting you a telegram I have just sent to the Managing Director:
>
> HUDSON FYSH ESQ
>
> QANTAS, SYDNEY
>
> SERVICE 1Q51 LANDED AT NEDLANDS ON SCHEDULE AT 0330 GMT TODAY AFTER 26 FLYING HOURS WITH CAPTAIN TAPP IN COMMAND STOP. THIS ARRIVAL MARKS THE COMPLETION OF THE HUNDREDTH CROSSING REPRESENTING 360,000 MILES BETWEEN THE TWO POINTS WITH REGULAR AIRLINE DEPENDABILITY STOP. CAPTAIN RITCHIE ALREADY PROCEEDING WITH RIGEL STAR AND NOW IN FLIGHT WITH THE ONE HUNDRED AND FIRST TRIP. CROWTHER

I would like to take this opportunity of congratulating the entire staff of this Division, who have, each one of you, contributed to this successful operational record. It is felt that this record had been unique, highly creditable and satisfying.

Passenger Experiences

Just what was it like for a passenger on the "Double Sunrise" service flying across the Indian Ocean in wartime?

Passengers boarded through the open rear fuselage observation blisters. The Catalina cabin had five separate compartments with a bulkhead door between each compartment - if the hull was holed, the door would be shut to restrict water to that one compartment. Neither crew nor passengers could move around the cabin for the first three hours after departure, when the fuel load made the aircraft tail heavy and the controls sensitive to changes in the centre of gravity. After three or four hours as fuel was burnt and the aircraft's weight decreased, the centre of gravity moved sufficiently forward that passengers were allowed out of their seats. They could use the toilet, make themselves a warm drink or move to the rear of the blisters to look out.

There are very few recorded passenger comments on the Indian Ocean Service. QEA Managing Director Hudson Fysh's personal diary covered his Ceylon-Perth crossing on Service 1Q8 30-31 August 1943 on board G-AGFL *Vega Star*:

> 8.30am: We are at 2,400 feet and our airspeed is 105 knots. Captain Crowther says it will be eleven hours before the aircraft has a single-engined performance. This knowledge, plus the fact that we have no dump valves, plus the sea below, plus the fact that we are flying into Japanese patrolled areas, lends a spice of risk to this trip

11am: At 5,500 feet doing 121 knots climbing to try for better winds.

12.40pm: While up forward with the crew scanning the ocean … a large ship lay directly in our path. It showed no white wake behind and was evidently not moving. Captain Crowther wisely sheared off and gave the ship a wide berth because it could have been an enemy raider.

5pm: The navigator, Flight Lieutenant Cowan, has just dropped a smoke flare to get a sight on to ascertain drift. We watched it go down and down, out of sight, then to finally burst on the water into a long stream of smoke. It showed a strong headwind still.

12 midnight: The stars are shining bright, and Crowther and Cowan have just taken a shot on Vega. Canopus is very bright and twinkling. Mars gleams steadily and I am told he is a most useful being, hung in the midnight firmament at right angles to our course … the handle of the saucepan is hidden by the big wing of the Catalina.

2am: Have snatched a little sleep. The crew seem indefatigable and are going strong.

5.15am: Crowther has a lie down. He has now been on duty for 24 hours. Several of the flame floats used during the night for navigation purposes were duds.

8am: Just had breakfast. 24 four hours in the air and there is the sun greeting us again as it rises in the east.

2pm (Colombo time): This makes 30 hours in the air. We are at 8,000 feet and it is as cold as hell.

3pm: Nose down for the Swan River in Perth and all changing into shore clothes.

3.45pm: We touch down on the Swan River 31 hours 45 minutes out from Koggala.

That flight's navigator, Jim Cowan, adds more detail to the mystery ship they turned to avoid. He had visited the RAF Group Operations Room in Ceylon the previous day and was aware they were searching for a disabled Allied fuel tanker, which had been torpedoed. The fuel it carried was desperately needed in Ceylon. Crowther agreed to allow the radio operator to transmit a morse message to RAF Ceylon, coded by the SYKO cipher machine, advising the accurate position of their sighting. The tanker was subsequently towed to Ceylon and the fuel salvaged. Cowan also said Hudson Fysh never forgot that flight, 32 hours buoy to buoy:

> When I saw him at various Qantas functions up to the Boeing 707 years, old Huddy always used to come across and say "Nothing like the Catalina days, Cowan?" and I would say "No, sir".

Catalina First Officer Ivan Peirce recalled a particularly arduous flight for one passenger:

> We had an American lieutenant on board with his bag of diplomatic mail strapped to his body in the usual American way - if you wanted to steal the bag, you had to chop their arm off. Five or six hours out from Perth heading for Ceylon we flew into a cyclone before we could warn him. The sea was stirred up, and we thought that something was brewing but the ride became rough, and we worked out we were in a cyclone which wasn't on the weather forecast. The sea was white and cream, and the navigator worked out that we were doing 253 knots, which was a colossal ground speed for a Catalina that was only doing 105 knots air speed. The skipper was John Shields, ex-RAAF, but attached to the RAF in England who had flown Catalinas and Sunderlands in terrible weather. He made a sharp 90 degree turn to port, to get us out of the cyclone. The poor old Cat was now only doing 15 knots ground speed for an hour because of the headwinds, but in the second hour we managed to get it up to around about 90 knots. We were going up with our nose down and down with our nose up and it was quite rough. It took five hours before we got right out of that cyclone. I felt sorry for our American passenger on his own down the back, all he could see out a small window was the wing flexing.

For passengers the flights were an ordeal, aboard a shatteringly noisy and unheated Catalina. Each ocean crossing was long, cold and lonely, mostly flown in darkness with many associated dangers. Because of the additional fuel tanks installed inside the aircraft, smoking was absolutely forbidden. Passengers would have been crucially aware that their wellbeing rested on the ability of the crew and the two marvellous Pratt and Whitney engines to get them to their destination. It must have been a sobering thought during such a long flight.

Catalinas Retired

Replaced by Liberators, the Catalinas would be withdrawn from service in mid-1945. With their high flying hours, maintenance was a growing problem, especially in respect to ever-present airframe corrosion. Final Catalina services took place during July 1945. The honour of the final service went to G-AGID *Rigel Star* arriving at Nedlands on 24 July as service 1Q135 flown by Captain John Solly. Also arriving at Nedlands that same day was G-AGFM *Altair Star* which had been dispatched on 20 July as a "Special Flight" Perth-Koggala-Perth: its purpose was probably to retrieve Qantas personnel and equipment from the Koggala Lake Qantas station, which was now closed.

Arrival dates at Nedlands for each aircraft's final Ceylon-Perth flights are listed in the table below:

The fifth Catalina G-AGKS *Spica Star* had been ferried Perth-Sydney on 25 March 1945 ahead of its CofA expiry due on 1 May. The intended CofA renewal overhaul at Rose Bay flying boat base was deferred because Indian Ocean Catalina services were being reduced in favour of the Liberators. Instead, G-AGKS was stored at Rose Bay and held in reserve.

The four Catalinas retired at Nedlands were parked on beaching gear and sealed for storage, awaiting instructions regarding their return to BOAC. They were left behind a locked gate when Qantas vacated the Nedlands base, removing all company facilities and shifting maintenance equipment across the Perth suburbs to RAAF Guildford, where Qantas had a hangar to service the Liberators.

Catalina aircrew and senior members of Qantas staff, together with a number of the company's associates who had been connected with the Catalina operation since its inception, were guests at a QEA function held at the Esplanade Hotel, Perth. Qantas Western Operations Division's Perth activities were now centred at RAAF Guildford, the military airfield six miles east of city. Before being posted elsewhere in the Qantas organisation, the Nedlands engineering staff held their own farewell party on 3 August 1945, sending out this invitation:

G-AGKS	Spica Star	10 Jan 1945	1Q107	Ferried from Nedlands to Rose Bay 25 March 1945, retired at Rose Bay
G-AGFL	Vega Star	9 July 1945	1Q132	Retired Nedlands
G-AGFM	Altair Star	18 July 1945	1Q134	
G-AGFM	Altair Star	24 July 1945	Special	Perth-Koggala-Perth. Then retired Nedlands
G-AGIE	Antares Star	13 July 1945	1Q133	Retired Nedlands
G-AGID	Rigel Star	24 July 1945	1Q135	Retired Nedlands

Fitted with beaching gear, a Catlina is towed up the Nedlands slipway for maintenance at the nearby hangar. (Qantas)

The Catalina Lament

Have you ever heard of Nedlands
the Catalina Base,
And the men who keep 'em flying
however hot the pace
The chaps who toil, and sweat
and swear and laugh
The men who keep employed
the salaried office staff
The bosses by the dozen, with uniform
and swank,
Who wouldn't know a piston from a
split pin or a crank.
Hear Ye! Hear Ye! O receiver
of this card.
The humble folk at Nedlands desire
you as their guest
At 4pm on Friday, where the Cats
are now at rest.
For the Catalina days are over, in this
State of WA,
And we wish to hold a party, to speed
them on their way.
So forget your mother's meetings, and
enjoy a jug of beer,
With the common folk at Nedlands, who
will make you welcome here.
The Staff, Nedlands.

What happened to these magnificent flying boats which had performed so faithfully, and upon which so much care and maintenance had been lavished? When BOAC advised the British Air Ministry that the five Catalinas had been retired in Australia, they were directed to hand them over to No. 300 Wing, RAF, then based in Sydney. Qantas was instructed to have them beached and stored on land pending a final decision as to their disposition: four at Nedlands and G-AGKS *Spica Star* at Rose Bay.

The Catalinas had been supplied to Britain under the terms of the United States government Lend-Lease Act, which allowed the US to lend or lease, rather than sell, war equipment to countries deemed vital to the defence of the United States. The Second Word War officially ended on 2 September 1945 following Japan's formal surrender. Two weeks later the US government terminated the lend-lease agreements, thereby requiring every item supplied under lend-lease to be paid for, or certified as destroyed in which case payment was waived.

Neither the RAF nor RAAF had any need for these early model non-amphibious Catalinas, which were therefore listed to be destroyed. No. 300 Wing issued orders that they be towed out to sea and scuttled, which at an administrative level, was probably seen as quick and easy disposal method. At that time the Royal Navy Fleet Air Arm in Australia was making arrangements for their aircraft carriers to take hundreds of its aircraft out to sea off Sydney and Maryborough, Queensland, to be pushed off the deck into the sea.

The four Catalinas at Nedlands became a complication when it was realised that the Fremantle Road Bridge would stop them being towed down the Swan River to the sea. The RAF destruction order was deftly changed to require the Nedlands Catalinas to be flown out to sea, rather than towed. The task was allocated to the Royal Australian Air Force.

Most published references quote November 1945 as the scuttling date for the Nedlands Catalinas. This was probably based on the date their British civil registrations were cancelled, as on 28 November 1945 they were noted as "Taken over by Australian authorities". However, the actual dates have been confirmed from the logbook of Arthur Jones, one of the RAAF airmen involved in the Nedlands scuttling operation (his full account is given in Appendix 4):

17 January 1946	G-AGIE	*Antares Star*
4 February 1946	G-AGID	*Rigel Star*
14 February 1946	G-AGFL	*Vega Star*
27 February 1946	G-AGFM	*Altair Star*

The fifth Catalina, the troublesome G-AGKS *Spica Star*, was retired at Rose Bay in March 1945. It met the same fate and was scuttled out to sea off Sydney. No specific date has been found, with most references giving March 1946, which is again probably based on the British civil registration cancellation date of 12 March 1946. However, an unexpected entry in Captain Orm Denny's logbook throws further light on her final days. Denny had flown for Qantas since 1936 but had not been sent to Perth, staying on Lockheed 10 and Empire flying boat operations through the war. His log for 6 November 1945 records three flights in G-AGKS at Rose Bay, "practice" then "Rathmines return, special flight."

The Catalina's CofA had expired on 1 May 1945, so some effort must have been expended to put the aircraft into flying condition. With no commercial value for Qantas, it can only be assumed this was carried out under the same No. 300 Wing order as the disposal of the four QEA Catalinas at Perth. This suggests that the scuttling date for *Spica Star* was probably in early November 1945.

The *Order of the Double Sunrise* Catalina era ended in July 1945, replaced by the *Order of the Longest Hop* Liberators and later the even faster Lancastrians.

Catalinas were to enjoy a second career with Qantas. Late in 1947 the airline acquired the first of a total of fourteen RAAF disposal PB2B-2 models with a taller tail fin, of which six were converted at Rose Bay to civil standards. These latter-day QEA Catalinas operated passenger services to Fiji, New Caledonia, the Solomon Islands, Lord Howe Island and within Papua New Guinea. The last two were retired in August 1958.

Although their domain was the South Pacific, one Qantas PB2B-2, VH-EBA, visited the Cocos Islands in May-June 1951 when "Double Sunrise" Catalina veteran Captain Len Grey carried a party of RAAF and DCA airfield construction officials to Cocos to assess the scope of work to rebuild the abandoned wartime RAF strip for use by Qantas Constellations. Because aviation fuel was not available at Cocos, the route went the long way around: Sydney (Rose Bay)-Brisbane-Darwin-Djakarta-Cocos-Djakarta-Australia, to allow fuel to be taken on at Djakarta for the return flight to Cocos.

G-AGID Rigel Star at sea off Rottnest Island on 4 February 1946 just prior to it being scuttled. (State Library of WA)

Chapter 4
Liberators: The Kangaroo Service

By early 1944 Qantas renewed its request that Consolidated Liberators should be released to provide more capacity on the Indian Ocean service. The four-engined Liberator had been the initial preference for the service because it offered a much higher payload and performance over the Catalinas. BOAC had been issued with former RAF Liberator bombers converted to unarmed passenger transports for some wartime routes to Cairo and Africa. BOAC also operated Liberators on the North Atlantic Return Ferry Service, returning ferry crews to collect their next American built bombers to be delivered across the Atlantic to Britain.

Qantas overtures intensified during an Australian visit by the BOAC chairman Lord Knollys during March 1944. Soon after his return to London he cabled Fysh to advise that two Liberators would be delivered to Qantas within the next two months. Captain Ambrose wrote in a company report:

> The news that our Catalina fleet was to be supplemented by two converted Liberators was received enthusiastically throughout the division. Great interest centred on Guildford aerodrome when during the first week of June 1944 the first LB-30 G-AGKT arrived under the command of Captain OP Jones, a veteran English airline skipper.

The introduction of land-based aircraft required QEA Western Operations Division to move part of its operations from the Nedlands marine base to an aerodrome. The existing Perth civil aerodrome at Maylands lacked sealed runways and was quite unsuitable for heavily laden Liberator departures. Given the wartime importance of the Indian Ocean service, Qantas was given approval to use the recently constructed RAAF station at Guildford, only a few miles from Maylands on the site of the pre-war Dunreath golf course.

At Guildford the Department of Civil Aviation had built a large hangar for civil airline use, with a wide aircraft parking tarmac and sealed taxiway. It was in an area cleared from scrub some distance away from the RAAF hangars. QEA had to establish new maintenance facilities at Guildford, while at the same time maintaining the existing Nedlands engineering for the Catalinas. The same irregular supply of essential parts, as experienced with the Catalinas was repeated with the Liberators. Aircraft instruments were particularly difficult to obtain, and it was not until the end of the war that the spares situation was resolved satisfactorily.

The Catalinas had been flying the Indian Ocean service for almost a year when the first Liberator was ferried to Perth by BOAC's famed senior commander Captain OP Jones, who was to provide operational and technical advice on the new type. To reduce the length of the ocean crossing, flights would refuel at Exmouth Gulf military airfield, 800 miles north of Perth. This new route would reduce the ocean crossing from the Catalina's 3,513 miles to 3,077 miles, and the higher cruising speed of the Liberator would cut at least ten hours from the flying time between Perth and Ceylon. Initial Qantas crews were drawn from those operating the Catalina service and were later augmented by Qantas staff transferred to Perth from other duties and seconded RAAF aircrew.

Some Catalina pilots had no experience flying aircraft equipped with a retractable undercarriage; their Qantas careers having been on DH.86s with a fixed undercarriage, Empire flying boats and Catalinas. The first QEA Liberator crews received conversions on RAAF B-24s from No. 7 Operational Training Unit at Tocumwal, New South Wales. For the following airline aircrews, a more formalised classroom and flying training course on the Liberator and its systems was necessary. Because the Indian Ocean Liberators could not be spared for such duties, approval was granted for a civilian conversion course for QEA pilots and flight engineers at No. 7 OTU, Tocumwal.

RAAF Flying Officer Ed Crabtree was a No. 7 OTU Liberator instructor, following a B-24 operational tour attached to the 530th Bombardment Squadron, USAAF, in the Darwin area. He recalled:

> Our pupils were mostly people who had flown in England or the Middle East on Lancasters, Halifaxes, Sunderlands, highly decorated for the most part. Tocumwal was noted for dust storms in those days with visibility reduced to about 50 yards. I remember one evening four of us were training pilots in night flying doing circuits when a dust storm went through. Two pilots diverted, one to Ballarat, the other to Laverton, leaving Bill Rehfisch and myself to have a go at landing. Bill and I contacted each other and found that we were both crossing the strip from different directions at the same time. Very hairy! We then gave one another turns at trying to get in. We both made it.
>
> Wing Commander Brill was later the CO; he had two tours on Lancasters and a later one leading the Pathfinders. An extraordinary man; very considerate, with a fantastic memory. In his early days at Tocumwal he lectured the instructors and said that we were training a very talented group of men who tended to become over-confident. So, his suggestion was to pull a motor on take-off, and if that did not do it, pull another one.
>
> As Qantas was getting converted Liberators for the Perth-Colombo run it was decided to check the crews through Tocumwal. I drew Captains Pollock and Hoskins, and believe me, they were extraordinarily skilled people; their handling of the aircraft was magnificent. Peter Hoskins joined us post-war in Trans-Australia Airlines and became senior route captain in Sydney.

QEA Western Operations Division manager Captain Crowther wrote to the DCA on 10 June 1944 detailing the Liberator's performance and the proposed QEA operation from Perth. He stated that the MAUW was 57,000 pounds, which included a maximum payload of 3,000 pounds: equivalent to eight passengers, their baggage and 320 pounds of mail. He advised that QEA had requested through BOAC to have the British CofA amended to allow a MAUW of 58,000 pounds. Nothing further is recorded in the DCA file, but from the inaugural Liberator service a week later, Qantas applied a 60,000 pounds MAUW, probably based on the forceful opinion of BOAC's Captain Jones.

The Liberators

Four Liberators were loaned to Qantas by BOAC for the Indian Ocean service, although the last aircraft was so delayed that it made only a single crossing, that of its delivery flight from England to Sydney. All were Consolidated Model 32-3 bombers, designed as LB-30Bs, built at the same Consolidated Aircraft plant at San Diego, California, as the Catalinas. They were purchased by Britain prior to Lend-Lease and delivered to the RAF as Liberator B Mk.IIs.

Registration	c/n	RAF Serial	CofA issued	Delivered to QEA	Fate
G-AGKT	117	AL619	21.5.44	3.6.44	Broken up Sydney Airport 12.47
G-AGKU	45	AL547	24.7.44	14.8.44	Broken up Sydney Airport 12.47
G-AGTI	39	AL541	28.11.45	4.12.45	to QEA VH-EAI 6.47, Broken up Sydney Airport 8.50
G-AGTJ	22	AL524	7.3.46	7.3.46	to QEA VH-EAJ 4.47, Broken up Sydney Airport 11.50

Their individual histories prior to Qantas were as follows:

G-AGKT: AL619 delivered to RAF in Britain 17 May 1942, issued to No. 1653 Conversion Unit at RAF Polebrook as a crew trainer, damaged 14 September 1942, repaired by Scottish Aviation Ltd and conversion to passenger transport completed 25.4.44, issued No. 511 Squadron but diverted to BOAC. Registered G-AGKT 9 May 1944 to BOAC, CofA issued 21 May 1944, delivered to Australia for QEA.

Passenger seats inside the fuselage of Qantas Liberator G-AGKT. (Qantas)

Loading baggage inside the nose compartment of Qantas Liberator G-AGKT at Guildford in 1944. (Qantas)

G-AGKU: AL547 delivered to RAF in Britain 22 October 1941, to No. 511 Squadron at RAF Lyneham, to Scottish Aviation Ltd for conversion to passenger transport, completed December 1942. Damaged on flight to Gibraltar 10 January 1943 (date unconfirmed); delivered to Scottish Aviation Ltd at Prestwick 9 December 1943, completed 22 June 1944. Registered G-AGKU 28 June 1944 to BOAC, CofA issued 25 July 1944, delivered to Australia for QEA.

G-AGTI: AL541 delivered to RAF in Britain 3 November 1941, issued No. 1445 (Ferry Training) Flight at RAF Lyneham, to No. 1653 Conversion Unit at RAF Polebrook as a crew trainer, back to No. 1445 Flight, to No. 301 Ferry Training Unit, Lyneham, issued 30 April 1943 to No. 159 Squadron in India as a transport. Issued to Scottish Aviation Ltd at Prestwick 6 April 1944 for overhaul and conversion to passenger transport for BOAC. Registered G-AGTI 25 September 1945 for BOAC, CofA 28 November, delivered to Australia for QEA.

G-AGTJ: AL524 delivered to RAF in Britain 14 December 1941, issued to No. 224 Squadron, RAF St. Eval for anti-submarine work, transferred to Coastal Command No. 120 Squadron at RAF Aldergrove, Northern Ireland, and Keflavik, Iceland, on Atlantic anti-submarine patrols. The squadron's Liberators attacked and sank fourteen German U-boats plus eight damaged; to No. 1445 (Ferry Training) Flight at RAF Lyneham, then No. 178 Squadron in Egypt on bombing and supply air drops in the Middle East; transferred to Transport Command in Middle East 27 April 1944, to Scottish Aviation Ltd, Prestwick 23 May 1944 for overhaul and conversion to passenger transport for BOAC. Registered G-AGTJ 25 September 1945, CofA issued 7 March 1946, handed over same day at Scottish Aviation Ltd to QEA Captain LR Ambrose, who departed on the ferry flight to Sydney.

BOAC had first used RAF Liberators on the North Atlantic Return Ferry Service, carrying

Camouflaged Liberator G-AGKT shortly after arriving at Guildford on 3 June 1944. (Qantas)

The second Qantas Liberator, G-AGKU, at Guildford in natural metal finish in 1945. (Qantas)

military aircrew non-stop between Prestwick, Scotland and Montreal, Quebec, as part of the mass delivery of US built military aircraft to the European theatre. A spartan conversion involved covering over the bomb bay with flooring and fitting oxygen tanks. Up to twenty men, wearing their warmest flying gear, could be carried lying on mattresses in sleeping bags with individual rubber tubes allowing them to breath oxygen when at high altitudes during the average fifteen-hour flight.

RAF Liberators of various models were also released to BOAC to maintain essential long-distance routes to Cairo and Africa. The four assigned to BOAC for use by Qantas were sent to Scottish Aviation Ltd at Prestwick Airport, Scotland, a major civil contractor to the RAF. They were significantly modified for passenger use. All gun positions and armament were removed, the bomb bay sealed, and the nose redesigned to include a freight compartment with a hinged nose cap. The passenger compartment, with sound-proofed walls, was in two sections: above the bomb bay and in the aft fuselage. Some were further refined with windows cut into the fuselage sides. The engines were the same reliable 1,200 hp Pratt & Whitney R-1830 Twin-Wasps as used on the Catalinas already in service with Qantas, although different models meant that engines were not interchangeable between the two types.

Liberator Service Begins

The first Liberator to arrive at Perth was G-AGKT, which landed at Guildford on 3 June 1944 at 1237 at the end of a ten-day delivery flight from England under the command of senior BOAC Captain OP Jones. With a flying career going back to the First World War, Jones had pioneered Imperial Airways empire routes and was credited with making the North Atlantic Return Ferry Service a viable wartime necessity. The strictly formal Captain Jones, whose fearsome reputation had preceded him, was to be the initial instructor in all aspects of the Liberator's operation.

G-AGKT's logbook indicated it had minimal RAF use, with only 285 hours 9 minutes flying time since manufacture. It was painted in Dark Green and Dark Earth camouflage over all upper surfaces and black underneath. The civil registration letters were black, outlined in white and the letters were underlined with the horizontal White and Blue (South East Asia Theatre) civilian nationality stripes. Blue and white RAF fin flashes were retained over the camouflage. The aircraft was soon engaged in crew training at Guildford under the severe guidance of Captain OP Jones.

By the time the first Liberator arrived, five captains and three first officers had been checked out during initial quite basic training on RAAF B-24s at Tocumwal. Captain Ambrose recalls his Liberator endorsement consisted of three daytime circuits and two nighttime circuits. Prior to leaving on the first trip to Ceylon, he flew a single circuit in G-AGKT at Guildford. Ambrose found the aircraft easy to handle, commenting that handling on and off the ground was similar to a flying boat.

During a local test flight for G-AGKT at Perth on 15 June 1944, the starboard dinghy-hatch, and the dinghy, were lost. The Liberator main wing spar was continuous and passed through the fuselage aft of the cockpit. Inside the cabin an overnight case had been stowed on the top of the spar and during take-off slid rearwards and stretched the dinghy release cables; the hatch flew off, and the dinghy promptly followed, trailing in the slipstream. A few seconds later the dinghy automatically inflated, its restraining rope broke, and the fully inflated dinghy tumbled gently to earth. Had the Liberator not had twin fins and rudders, the dinghy almost certainly would have wrapped itself around a single fin and caused loss of directional control. A new hatch was quickly fabricated and the aircraft entered service on the Indian Ocean route two days later.

The first Liberator service was scheduled for 17 June 1944. It came close to being postponed due to an incident on the ground at Perth during the pre-departure test flight. The BOAC first officer had diligently commenced his pre-flight checklist prior to Captain Jones boarding the aircraft. When testing the flap control lever, he inadvertently selected the undercarriage retraction. The retraction sequence was nose wheel then main gear – when the aircraft's nose dipped slowly, he immediately realised his error and reset the lever to "Landing Gear Down". Fortunately, there was sufficient hydraulic back pressure to extend the nose gear back up and the undercarriage indicator light gave the "Landing Gear Down and Locked" indication. No damage had occurred and he hoped nobody had seen the minor nose dip. However, Captain Jones was in the despatch office checking the load sheet and through the open door saw the nose drop then rise back to normal. Jones took his left seat in the cockpit and conducted the test flight without any mention that he had seen the event. Some hours later during the actual service, Jones waited until the QEA pilots had left the cockpit to give a scathing reprimand to his first officer in the noisy privacy of the cockpit.

The inaugural Liberator Service 4Q1 departed Guildford on 17 June 1944 under the command of Captain Jones with his BOAC ferry crew, but carrying Qantas Captains Ambrose and Crowther and engineer Roberts for route training. The flight took three hours 40 minutes to reach the new stopping point at Exmouth Gulf. After refuelling, G-AGKT used the full length of the runway to become airborne, even with the cool southern winter temperatures and its all-up-weight reduced from the maximum 60,000 pounds to 56,000 pounds for the first trip.

Colombo was reached after a 16 hour 13 minute ocean crossing, the Liberator landing at Ratmalana aerodrome after an uneventful flight. The return trip with the same crew, except that Captain RJ Ritchie replaced Captain Crowther, left Ceylon on 21 June, taking 17 hours 21 minutes flying time to Exmouth Gulf. The next day the flying time to Perth was three hours 50 minutes.

After the first service, Jones made it clear he was concerned about the Qantas pilots' apparent lack of discipline and informal manner, such as the crew addressing captains by their Christian names in conversation. However, QEA's subsequent accident-free Liberator operations, improving take-off performance and load-carrying figures, proved him wrong. Ambrose was appointed Perth training captain to check out company pilots who had completed the RAAF conversion training at Tocumwal.

The next Liberator service 4Q2 left Perth on 28 June, again commanded by BOAC Captain Jones with Qantas First Officer JW Solly. From now the Liberator schedule was planned to be a Perth departure every ten days, manned by Qantas crews.

Qantas' second Liberator, G-AGKU, arrived in Perth on 14 August 1944 following a fifteen-day ferry from England by a BOAC crew, with delays due to hydraulic problems and crew illness. The aircraft was delivered in natural metal finish, with nationality stripes under the registration letters and fin flashes. On 30 August it was flown from Perth to Archerfield aerodrome, Brisbane, for modifications. The QEA Archerfield workshops had extensive experience with Liberators because of wartime contracts to repair, service and modify USAAF B-24 Liberators. Work carried out on G-AGKU included the removal of auxiliary fuel tanks from the bomb bay and installing replacement rubber fuel cells inside the mainplane. This provided additional bomb bay space for passenger baggage and mail sacks, allowing QEA Liberators to carry fifteen passengers and five crew. This was a total payload of 5,500 pounds compared with the Catalinas' meagre 1,000 pounds payload.

When work on G-AGKU was completed at Archerfield it was test flown on 19 September, before departing two days later for RAAF Laverton near Melbourne. At Laverton No.1 Aircraft Depot had weighing scales suitable for the Liberator. Weighing was necessary to establish empty weight and the precise centre of gravity, both altered by the modifications. A recently employed Qantas engineer, Ron J Yates, was the Qantas representative during the Laverton weighing. Ron Yates remembered it as one of his first jobs with Qantas following his transfer from RAAF service as an aeronautical engineer. He was later to rise to the position of chief executive officer of Qantas Airways Ltd. G-AGKU positioned from Laverton to Perth on 23 September in time for its first scheduled Perth-Ceylon service 4Q11 on 24 September. This allowed the original Liberator to go to Archerfield for similar modifications and Captain Ambrose ferried G-AGKT to Brisbane on 19 September 1944.

Unfortunately, G-AGKU's Indian Ocean service was soon interrupted when the nose wheel assembly collapsed during landing at Guildford on 16 October 1944. Earlier that day Captain Tapp had attempted to depart on service 4Q14 to Ceylon, but faults caused the flight to be postponed while problems were rectified, and the aircraft twice test flown. Late that afternoon Captain Tapp was carrying out a third test flight with crew manning their positions, each with their own checklists. The crew was First Officer JA Furze, Second Officer RJ Fairservice (under training), Navigation Officer HJ Bartsch and Radio Officer WR Clarke.

Liberator G-AGKU following its nose gear collapse at Guildford on 16 October 1944. (Qantas)

The nose of Liberator G-AGKU under repair at Guildford. (Norm Roberts)

The senior station engineer Roberts was on board, with electrical engineer L Weekes to check the electrical system failures. In addition, three ground staff were seated in the rear passenger seats, tasked with distributing the already loaded freight and mail between the nose and main cargo compartments to affect a correct Centre of Gravity for the coming service. After 37 minutes Tapp returned to Guildford but when the undercarriage was lowered, the cockpit indicator light showed the nose gear was not locked. These indicators had proven to be unreliable, so Tapp instructed Roberts to go down into the nose compartment, lie on his back and kick the nose wheel lock into position with his boot. When Roberts reported back that he could not be sure it was locked, Tapp continued for a landing, throttles fully back crossing the fence, touching down at 90mph and keeping the nose wheel in the air. When the nose settled the wheel folded and the aircraft nose dragged along the centre of the runway. Captain Tapp's official accident report to Qantas and the DCA included the following candid assessment of this second Liberator sent to them by BOAC:

> This is the latest act of many troubles experienced with this aircraft, only some of which seem to have any logic attached to them. Owing to the life history of this aircraft, I consider it needs another complete and very thorough examination including all its accessories before it again goes into service. In its present state, it gives me a feeling of no confidence in the job it has to do.

Back at the hangar Roberts determined that the nose section was beyond repair. A distorted frame had been preventing free alignment of the undercarriage link mechanism. Roberts' enquiries revealed that the undercarriage frame damage had been sustained by an earlier nose wheel collapse while in RAF service in North Africa. He needed a full nose gear replacement in a hurry, and he knew where to find it.

Roberts flew to Brisbane where the Qantas workshops at Archerfield aerodrome had a collection of wrecked and discarded USAAF B-24s from their military servicing contract. He arranged to have the nose gear structure cut from a damaged B-24 being stripped for parts by Qantas. The replacement section was flown to Perth in a USAAF C-47 then operated by QEA on military courier runs to New Guinea. Roberts supervised the nose repair in the Perth hangar, which was completed by 11 November 1944, when Captain Ambrose took G-AGKU on an extended test flight to thoroughly check every system in the troublesome Liberator. This involved flying across the Australian continent to Melbourne and back, to collect an urgently needed item. He departed Perth at 1940 that evening for a night astronavigation crossing, before landing at Essendon aerodrome, Melbourne, at 0620 the next morning. After only two hours on the ground, Ambrose was airborne again and arrived back at Perth at 1742 that afternoon. Flight times were eight hours 39 minutes to Melbourne, and eight hours 56 minutes on the return. Next morning G-AGKU departed Perth on schedule, with Captain Tapp operating service 4Q19 to Ceylon.

By January 1945 the Liberators were making four crossings every week. Service 3Q100 Colombo-Perth overnight 26-27 August 1945 was the 200th Liberator Indian Ocean crossing. One month later the 2,000th Indian Ocean passenger was carried.

Kangaroo Service

Soon after the first Liberator arrived in Perth, Captain Crowther was visiting Sydney to discuss the new operation with senior Qantas management. Hudson Fysh had made a casual suggestion that the Liberator service be given a name and suggested "Kangaroo Service" might be a good title. On the trip back to Perth, Crowther became quite enthusiastic about the idea, and, after trying the name on several of his men in Perth as well as Captain Jones of BOAC, he sent a telegram to Fysh officially requesting the new Liberator service be named the "Kangaroo Service".

The original 1944 kangaroo design used by QEA was the same as that on the Australian penny coin. Australia's armed services adopted that same design twelve years later when the hopping kangaroo was introduced as the centrepiece of aircraft nationality roundels. While the style of the kangaroo has changed over the years since, it became the symbol by which Qantas is now recognised around the world - the long hop across the Indian Ocean was to spawn a company symbol.

Meanwhile at Archerfield G-AGKT was test flown on 12 October after its modifications and overhaul. It emerged from the hangar with camouflage removed, revealing a gleaming metallic finish similar to its partner. Both aircraft were now in the same markings, consisting of British registrations in large black letters on the fuselage sides and wings; all lettering having the usual underlining blue and white nationality stripe. The only other concession to the wartime situation was the military blue and white fin flashes, painted either side of the tail fin. A dark blue nose and cheat-line ran down the fuselage sides, at the aft end of which was the red BOAC Speedbird emblem and the words "Trans Ocean". However, the most interesting new marking was under the cockpit: the now familiar leaping kangaroo made its first appearance on a company aircraft, together with the words "Qantas Empire Airways, Kangaroo Service".

G-AGKT departed Archerfield on 16 October 1944 at 1320, bound for Perth via a stop in Sydney for crew to attend to company business. Soon after arrival in Sydney a message was received from QEA Perth instructing them to proceed directly to Perth to replace Liberator G-AGKU damaged that day at Perth by the nose wheel collapse. G-AGKT departed Mascot on 18 October at 0936 for Perth and two days later resumed Indian Ocean flights as service 4Q15 under the command of Captain Ambrose.

When the two Liberators were both in service, their fifteen passenger seats were able to handle the brunt of passengers until the introduction of the Avro Lancastrians the following year. Company, crews and passengers were all to benefit from the ability of the Liberators to carry bigger payloads in a shorter time. The safety factor was also greatly enhanced by the Liberator's four engines, and their speed reduced the hours spent over the ocean. An additional very welcome bonus was provision for a flight steward and galley. The steward managed the passengers' requirements and refreshments, also delivering hot meals to the cockpit. The schedules were structured to make the greater part of each ocean crossing in darkness to allow for astronavigation.

When the Liberators entered service, Qantas was providing four services fortnightly in each direction across the Indian Ocean. Over the next year the Catalina frequency was gradually reduced, from three times per fortnight to once fortnightly, as the Liberators took over their services. Just as passengers on the Catalina service were presented with a certificate, Kangaroo Service passengers received flight certificates, signed by the captain, entitling them to membership to "The Elevated Order of the Longest Hop".

Parts and Spares Supply

The power plants of the QEA Liberators were not fitted with turbochargers as used by B-24 bombers. The engine cowlings of the QEA aircraft were circular instead of the usual distinctive B-24 oval-shape to accommodate turbocharger air scoops on either side of the engine. The Qantas machines had Curtiss Electric propellers instead of Hamilton Standards. These three-bladed electrically actuated constant-speed propellers were manufactured by Curtiss Wright under the name Curtiss Electric.

The Curtiss Electric propellers were notorious for "runaways", when a malfunction forced the blades into fine pitch, causing the propeller to windmill up to dangerously high RPM in the airflow. If unable to be feathered, bearings would seize, and the propeller unit could break away in flight. During July 1945 both G-AGKT and G-AGKU suffered runaway propellers soon after take-off. Replacing the electric hubs had initially failed to solve the problem. Milton Getker, a field representative for Curtiss Wright, made three trips to Perth from the forward battle areas to bring first-hand technical information and vital spare parts for the QEA Liberators' propellers. These technical visits, which helped Qantas to minimise propeller problems, were arranged through Ernest Heymanson in Melbourne, whose business EL Heymanson & Company was the Australian agent for Curtiss Wright.

On 18 December 1945 G-AGKU made an emergency landing at Exmouth Gulf after suffering a runaway propeller soon after departure. The engine suffered severe internal damage requiring an engine change in open weather at this remote airfield. The third and fourth Liberators were delivered to Qantas with Hamilton Standard propellers.

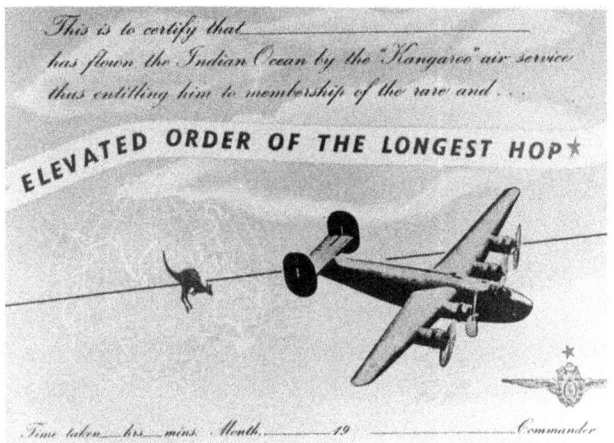

The Elevated Order of the Longest Hop certificate as issued to passengers using the Qantas Liberator Indian Ocean service.

From the beginning of the Indian Ocean operation, Qantas had used BG sparkplugs in the engines of their Catalinas and then Liberators. During the war these sparkplugs were manufactured in Australia by Leggetts Pty Ltd, Melbourne, under licence to the American BG Corporation. Leggetts manager Laurie Brown made several rushed visits to Perth to personally ensure occasional problems with his Australian product were promptly solved.

Qantas was not provided with any initial supply of spares for the Liberators, nor the essential technical manuals to provide guidance for maintenance. It was a frustrating situation. Perth senior station engineer Norm Roberts decided on a direct personal approach to American B-24 units then in the Darwin area. He arranged priority travel on the MacRobertson-Miller Aviation Lockheed 10 service from Perth to Batchelor, just south of Darwin. Roberts' intention to go straight to the Americans was frustrated when the RAAF area commander learnt of his mission and insisted that an RAAF stores officer accompany him.

The USAAF 380th Bombardment Group with B-24s was in the process of transferring base from the Darwin area to Biak in New Guinea. The move was nearly completed and many of their required spares were already crated awaiting shipment - but many unwanted items remained in their store. When told of Qantas' dire spares situation, the Americans invited Roberts to help himself from the remaining stocks. Norm Roberts recalls his elation as being like a child being let loose in the toy section of a department store. He obtained virtually every one of his requisites. Many were in component form although he did select some fully assembled units. His booty ultimately comprised half a Douglas load. It was a typical American wartime gesture and clearly intended to be free of charge and off the books.

At that time Qantas was operating USAAF Douglas C-47s and C-53s on military courier transport flights between Australia, New Guinea and Borneo. Roberts deftly arranged to have a Qantas operated C-47 and crew available in Darwin between rostered courier runs, to fly his parts directly to Perth. However, it was not to be as the RAAF stores officer insisted that every item would have to be listed and indented from the USAAF to the RAAF, who would in turn charge Qantas. The Americans protested that this was unnecessary, but RAAF propriety had to be observed. As a result, the items urgently needed in Perth were despatched to Brisbane where the obligatory paperwork, to list and charge every part, took the RAAF stores section several weeks. They were then flown to Adelaide and finally reached Perth by an ANA DC-3, with Qantas having to pay a hefty ANA air freight bill. The Americans' generosity, with spares they did not need, thus incurred considerable expense and effort. Nevertheless, Roberts obtained his spares needed to keep the Liberators flying.

Perth (Guildford)

The Australian terminal for the new landplane service was Perth's recently constructed airfield, RAAF Guildford, five miles from the city. This new military airfield was established to reduce the pressure on RAAF Pearce, fifteen miles to the north. Under wartime provisions the government had taken over the lands and golf course of Dunreath Estate and two sealed runways had been completed the previous year, when it was originally referred to as Dunreath aerodrome. Various RAAF units were based at Guildford for the last two years of the war including No. 85 Squadron with Spitfires.

The DCA had originally selected the site in 1938 for a new aerodrome capable of handling larger airliners, but the war in the Pacific resulted in the airfield being completed for the RAAF. Nevertheless, the DCA arranged for the construction of a civil airline hangar with taxiways and an aircraft parking apron remote from the RAAF hangar area. Australian National Airways was permitted to transfer from the smaller Maylands aerodrome to Guildford for their DC-2 and DC-3 services to Adelaide. When the QEA Liberators were introduced, DCA agreed to their use of the ANA hangar. Because the ANA reduced wartime services to Perth involved a Douglas night-stopping in Perth only two or three times weekly with no scheduled maintenance, Qantas engineering staff had the hangar mostly to themselves.

Liberator and Catalina passengers and mail travelling on to the Australian eastern states were transferred at Guildford to ANA DC-2 or DC-3 scheduled services to Adelaide and on to Sydney. At Guildford passengers used the newly built ANA lounge next to the Guildford hangar and ANA was contracted to provide catering for the Liberators at Guildford. Two meals were served on the ocean crossing, provided in cartons and kept in an ice-chest on board, followed by hot toast from toasters specially made in Perth.

By the time the QEA Indian Ocean route dropped Perth in favour of Learmonth direct to Sydney, the RAAF had vacated Guildford, and the airfield was transferred to the DCA to become Perth Airport. It was not a designated Customs Entry Airport until 1952 when Qantas commenced Lockheed Constellation services Perth-Cocos Islands-Mauritius-Johannesburg. Since then, it has been named Perth International Airport.

Exmouth (Learmonth)

The RAAF airfield at Exmouth Gulf was constructed during 1942, originally at the request of the US Navy to provide fighter aircraft protection for an American submarine base at Exmouth Gulf. The airfield was given the wartime security code name "Potshot". Aircrew logbooks referred to the airfield by that code name, or Exmouth. In late 1944 the RAAF station was named Learmonth, in honour of Wing Commander Charles C Learmonth, DFC and Bar, who had commanded No. 14 Squadron at Pearce. He was killed on 6 January 1944 when his Bristol Beaufort crashed into the sea. At that time Australian-built Beauforts were suffering a high and unexplained accident rate. When Learmonth's aircraft began shaking violently, before completely losing control he radioed his wingman to inspect the tailplane, where an elevator trim tab control rod was seen to be detached. Subsequently all RAAF Beauforts had that trim tab modified, and Learmonth is credited with saving the lives of many Beaufort airmen.

After the war Learmonth became a civil airport for the town of Exmouth. From the 1950s, tensions between Indonesia and Australia resulted in the airfield being developed as an RAAF tactical forward "bare base" without resident units, to be used for deployments during a defence emergency. A civil passenger terminal and facilities were included as a joint-use airport, and today it is named RAAF Learmonth and Learmonth Airport.

Back in 1944 when Qantas received approval to refuel at Exmouth Gulf with the new Liberator service, the remote military base had few facilities to offer the airline. The captains' flight reports from early Liberator services confirmed BOAC training Captain OP Jones' contention that runway extensions at Exmouth Gulf were necessary. A deep ditch at one end of the runway was an additional hazard, so QEA aircrew were relieved when ground works began soon after. The completed 5,000-foot runway with a sealed surface proved satisfactory for the Liberators - and the Lancastrians which were to follow.

Strengthened cyclone-proof buildings were manufactured in Perth to a Qantas order. They were constructed from masonite and jarrah wood, with insulating canite to provide scant relief from Learmonth's high temperatures. These buildings were dismantled and shipped to Learmonth, where they were erected in the saltbush on the edge of the Liberator parking area. One building was the Qantas stores shed, outside which a chicken run was made in the scrub. An initial batch of 60 chickens was sent from Perth on an early Liberator service. This rather basic catering operation successfully provided eggs and meat for meals on the flights to Ceylon.

Another building became the passenger lounge. It was painted in light pastel shades, giving a cheerful contrast to the arid country outside. Qantas staff stationed at Learmonth initially had a jeep at their disposal until July 1945 when a Marmon Harrington three-ton truck was driven up from Perth. Company engineers took six days to make this overland journey on poor roads, the truck being fully loaded with Lancastrian spares and building panels. The truck was then stationed at Learmonth and was used to drive to the wharf at Exmouth for provisions and also as an aircraft tug.

Passengers from Liberator G-AGKT stretch their legs during a Learmonth stopover in 1945. (Geoff Goodall Collection)

Liberator G-AGKT flies the Australian Civil Aviation Ensign during refuelling at Learmonth. (Qantas)

Liberator and Catalina flight engineer positions were manned by Perth-based maintenance engineers endorsed for the flight engineer role. After the Liberators settled into service, flight engineers were not usually carried between Learmonth and Ceylon, because problems were anticipated to be rectified on the first leg from Perth. This allowed extra payload to be carried across the ocean but gave the engineer two days on the ground at Learmonth waiting for the aircraft's return, when he would be part of the crew for the flight back to Perth. These breaks at Learmonth were hardly a holiday, because there was little for the engineers to do, except fishing, and the RAAF mess provided only minimal hospitality towards the Qantas "civies".

From June 1945 Learmonth was used by both Liberators and Lancastrians, and with increasing numbers of passengers passing through, the QEA terminal's basic facilities were improved. The Qantas base shared the desolate airfield with a DCA Aeradio Unit and occasional RAAF detachments. By February 1946, Learmonth was handling ten Qantas flights weekly. But that was soon to end as Singapore and the Netherlands East Indies re-opened to airlines.

Also at Learmonth were Catalina detachments of No. 112 Air Sea Rescue Flight from Darwin, which from July 1945 provided search and rescue cover for the Qantas Indian Ocean flights. For this task, No. 112 Air Sea Rescue Flight also had detachments at the disused US Navy flying boat base at Crawley Bay, Perth.

Qantas Steward CF Baron's personal diary records his numerous crew rest stops at Learmonth while he was on Liberators and Lancastrians. Some examples:

- 1 September 1945. Lancastrian G-AGME. Captain Nicholl. Left Sydney 7.30am for Gawler, there for an hour then to Learmonth, landed 6.30pm. Only a RAAF base there, stayed three days – nothing to do but sleep and swim. Left Learmonth at 10pm on 4 September in G-AGMC for long ocean hop to Ceylon. Passengers included Minister for Foreign Affairs Dr Evatt and Mrs Evatt. He was very scared on this 14.5-hour flight to Ratmalana. I stayed two days at the Mount Lavinia Hotel near Colombo. Left for Karachi in G-AGLY. Engine trouble soon after departure so returned to Ratmalana but could not land for an hour because they were working on the runway. Away next day, 6.5 hours to Karachi, flying along the west coast of India. Karachi hasn't changed much since my Empire boat days.

- 10 December 1945: Lancastrian. Service running late, left Negombo at 8.30pm. Had Captain Scotty Allan as a passenger across the ocean. Landed at Learmonth at 1.30pm, very hot and dusty. Two days of dust storms, then to Sydney on Dec 14.

- 28 December 1945: Liberator G-AGTI, Captain Ritchie. Left Sydney for Gawler in the new Liberator, a good trip on to Learmonth. I was there for six days, cyclone hit the place, not much damage but the next service went via Perth. I left Learmonth on G-AGKT 3 January at 4.30pm for Negombo, landed there at 6.30am next morning.

A lyrical Qantas Lancastrian crewmember described Learmonth in early 1946:

A collection of prefab huts in the midst of a sand dune wilderness. Mile upon mile of virgin beach and warm green sea beyond – and flies ... the long lumbering take-off, the slow, ponderous climb-out with 3,000 miles of lonely sea ahead.

Ceylon

Qantas used RAF Ratmalana at Colombo for the Ceylon terminus. It had been Colombo's pre-war civil aerodrome and its terminal building facilities were adequate to handle passengers and freight for arrivals and departures. Qantas crew rest accommodation at Colombo was the Galle Face Hotel for pilots and the Mount Lavinia Hotel for the rest of the crew.

Ratmalana's single runway had inadequate length for the long take-off roll required by the heavily laden Qantas Liberators. Departures for Australia were required to ferry from Ratmalana with only a light fuel load to an RAF station with sufficient runway length for the heavy take-off. The initial RAF station nominated was RAF Sigiriya located 100 miles northeast of Colombo, the ferry taking an average flight time of 50 minutes. Although Sigiriya was a base for RAF maritime patrol bombers, the runway had mountainous terrain near one end and provided only the bare minimum required length for Qantas Liberator departures. The airfield was two miles from the ancient fortress city from which its name was derived. Several thousand military personnel were dispersed in the jungle around the airfield.

At first Ratmalana's runway was disturbingly short in one direction for Liberator take-offs, even

Colombo's Ratmalana terminal building in 1947. (Roger Thiedeman Collection)

though lightly laden for the ferry sector. A growth of tall palm trees at one end of the runway reduced the effective length for take-off. Qantas' requests to have the palms removed were rewarded when a team of elephants was brought in and the offending palms were soon felled.

Several months after the Liberators entered service, heavy departures were switched from Sigiriya to RAF Minneriya, twenty miles further to the north-east. Minneriya's runway was 7,500 feet long, which was more than adequate. Service 3Q6 flown by Captain JW Solly in G-AGKT made the first Minneriya departure on 21 August 1944, bound for Learmonth after a 54-minute ferry from Ratmalana.

Ratmalana runway extension works were carried out during 1945 but were still insufficient for the Qantas heavy departures for Indian Ocean crossings. Talks between Qantas and RAF Ceylon resulted in RAF Negombo being made the Ceylon terminus for both Liberators and Lancastrians. This military airfield had been constructed in 1940 in a coconut tree plantation near Colombo. With the decreasing military operations after VJ Day, its long runways and staging post facilities were made available to Qantas. Negombo allowed fully laden departures, ending the cumbersome positioning ferry from Ratmalana to a heavy departure airfield. QEA Liberators and Lancastrians commenced using Negombo from December 1945 and it continued as the only Qantas airfield in Ceylon until termination of the Indian Ocean route.

Colombo-Karachi Connection Problems

The British arrangements to introduce Liberators for Qantas covered only Australia-Ceylon, with no continuation to Karachi like for the Catalinas. The Liberators, which carried over four times the Catalina's payload, terminated at Colombo. Through passengers and mail faced a variety of RAF, BOAC and Indian airline connections between Ceylon and Karachi where they connected with the BOAC service to Great Britain. The QEA Operations Manager, Captain Lester Brain, set out this problem in a letter to BOAC dated 19 July 1944:

We now have one Liberator which has already operated several services between Perth and Ceylon. The second Liberator is expected to leave England next week and it is contemplated that on the arrival of this machine we will be able to operate a regular Liberator service once weekly each way between Perth and Ceylon. At this date there is no firm connection arranged between Colombo and Karachi which will tie in with the Liberator service and provide for the on-carriage of airgraph mails, passengers and other items of through-loading from Colombo to Karachi, whence regular and frequent services continue to the United Kingdom.

If it is found that the transit time between Australia and England is erratic and slower when using the new Liberator service (which was greeted in Australia as being faster and safer over the Indian Ocean stage) then the reaction will be one of disappointment and dissatisfaction. It is most desirable that once regular Liberator schedules are operating to Colombo, there should be a prompt connection by landplane on to Karachi and vice versa.

This frustrating gap appears to have occurred because the Air Ministry believed BOAC would be soon providing a landplane connection between Colombo and Karachi. BOAC had intended to operate such a service but attempts to use Armstrong Whitworth Ensign and Lockheed Lodestar aircraft already operating Karachi-Cairo were thwarted by aircraft losses. A May 1944 BOAC submission to the Air Ministry proposed a twice-weekly service England-Karachi-Colombo to link with the QEA Liberator service at Colombo. It would require five Liberator III transports to be loaned by the RAF, but that was refused due to military requirements. The air connection with Karachi remained unsatisfactory until early 1945 when the RAF introduced an Avro York courier service between Colombo and Karachi.

Former RAF Sunderland Mk.III flying boats were used by BOAC on the England-Karachi service. A first BOAC Sunderland run on the route took place on 11 October 1943 when G-AGET and G-AGEV departed Poole (Bournemouth) for

Karachi. Designated "Special Flights", they carried a party with the highest government priority status. Both aircraft had been prepared at BOAC's Hythe flying boat base, having VIP seating installed. G-AGEV also had replacement flame dampeners fitted to its engine exhausts. On the first day it was planned to reach Cairo via a refuelling stop at Djerba, Tunisia. However, G-AGEV turned back due weather, diverting to Pembroke dock. The following day G-AGEV staged via Gibraltar and Djerba to catch up at Cairo. On 16 October the pair continued to Bahrein and reached Karachi the next morning. After staying three days, the pair departed for England and reached Hythe on 27 October where the VIP seating was removed, and bench seating was installed.

The inaugural BOAC scheduled Sunderland service departed Poole on Christmas Day 1944 as service 19M1 operated by G-AGEV. On 2 January 1945 the same aircraft left Karachi on the first return service 20M1 to Poole. The BOAC Sunderlands were camouflaged and painted with RAF codes as well as British civil registrations because the service was operated under RAF Transport Command control.

Further improvement came in January 1945 when BOAC introduced Avro York landplanes on the England-Karachi route. On New Year's Day G-AGJD departed Hurn on the inaugural York service 27M37, reaching Karachi six days later. The same York left Karachi on 9 January 1945 on the first westbound York service, reaching Hurn on 13 January. Soon after BOAC Yorks began operating between England and Karachi, RAF Transport Command had sufficient transport aircraft to implement a scheduled Karachi-Colombo courier service with RAF Yorks. Qantas finally had an efficient connection for its passengers, freight and mail bound for Britain.

No Longer a Secret

The greater payload of the Liberators allowed the resumption of normal paper airmail between Great Britain and Australia, replacing the microfilm Airgraph Mail. Commencing in August 1944 paper letters in envelopes were once again allowed. An

Mail is loaded aboard Liberator G-AGKU at Guildford. International outbound paper airmail services were resumed in September 1944. (Geoff Goodall Collection)

innovation to reduce airmail postage cost was the Seven Pence Australian Air Letter forms, which went on sale at Australian post offices on 11 September 1944. The return to paper letters to Britain marked the beginning of a return to normal times for the war-isolated people of Australia.

The first inbound paper airmail from Britain was postmarked London on 24 August 1944 and arrived at Perth on 2 September on service 1Q82. Onward-bound letters reached Sydney two days later. The first outbound "paper" airmail left Perth by Liberator on 15 September 1944 for Ceylon and London. The event was marked by a ceremony at Sydney's General Post Office. Qantas used the occasion to make the first public announcement of its secret operations across the Indian Ocean. The company's statement revealed that the service had been in operation for over a year, and said in part:

> The Australian operated air service across the Indian Ocean has been of extreme importance in the successful prosecution of the war in the South Pacific area and has formed a vital link with other wartime transport systems.
>
> To say that the undertaking was hazardous is understating the facts. The route, which for the major part passes through Japanese patrolled areas, must be flown in complete radio silence and without the slightest protection of defence against enemy attack. Courses are plotted by astronavigation and dead reckoning, a science in which veteran Qantas captains are adept. The

remarkable success shown in the regularity of schedule is a tribute to the indomitable spirit which has always been manifest in Australian aviation and in Qantas operations in particular.

Finally, a sign went up at the entrance of the unassuming dirt road between trees leading to the Nedlands flying boat base reading "Qantas Empire Airways Ltd Marine Air Base". At Perth Airport, on Great Eastern Highway at the corner with Fauntleroy Avenue which led to the Qantas hangar, a much more imposing signboard was erected reading "Qantas Empire Airways Ltd – Western Operations Division – Guildford Airport".

The Liberators now settled down to their regular schedules, taking between sixteen and eighteen hours on each ocean crossing between Ceylon and Learmonth. Flying was mainly at night to allow for astronavigation. By development of cruising techniques and employing a Catalina-proven method of manually leaning the fuel-air mixtures to the engines, Qantas crews stretched the Liberators' range to 3,840 miles at an airspeed of 170mph. A crew duty roster pioneered on the Catalinas continued on Liberators, allowing each officer to have two breaks of two or three hours each. Bunks for the resting crew members were fitted at the end of the Liberator passenger compartment.

QEA Managing Director Hudson Fysh had been lobbying BOAC for additional newer Liberators to be released for the Indian Ocean service. His frustration is reflected in his diary entry of 27 October 1944 during seventeen hours on G-AGKT, crossing from Learmonth to Ceylon on his way to London for urgent talks:

> Our two Libs are the world's oldest and worst on the world's longest air hop.

Old they might have been, but Qantas flight crews found them to be good aircraft, which gave reliable service throughout their time on the Indian Ocean operation. They were operated by QEA to a maximum take-off weight of 60,000 pounds, which was 5,000 pounds overload. Decades later Captain Ambrose recalled:

> Our Liberators took to their two-ton overload as if they were born to it.

Although it was intended originally that all eastbound aircraft would land at Learmonth for refuelling, prevailing westerlies at altitude encouraged direct flights from Ceylon to Perth, providing the tail winds left an adequate fuel reserve on arrival at Perth.

The following account by a passenger on a typical eastbound crossing was published late in 1944, shortly after the operation was made public knowledge:

The Qantas Empire Airways sign erected at Guildford airport in 1944. (Qantas)

There is no runway near Colombo long enough to get a fully loaded Liberator safely off full. We took off from there with one thousand gallons of petrol and flew to a runway elsewhere in the island. There the fuel load was increased to 2,900 gallons – about 60 years' ration for a large car in Australia - while passengers and crew had lunch. Their next meal was to be breakfast in Western Australia.

We had with us on this trip Squadron Leader John Hogan of the RAAF in Perth, who was described as the wizard of Indian Ocean meteorology and whose calculations play an important part in deciding the exact course to be taken over the world's longest regular civil aviation non-stop route.

Normally the worst weather on the voyage is encountered around the equator when the aircraft strike what is known as the "inter-tropical front". It was a relief to be through the last of the bucketing squalls and eventually to be at a calm and comfortable height and stay there for about ten hours without a bump until we got home.

They took me up to the flight deck and dazzled me with the science of mathematics and navigation and with instruments and slide rules that work out the calculations so minute in themselves but so fundamental to safety on a flight that in peacetime would have been reckoned as adventure.

Captain Thomas was up there peering below his long eyeshade. He has made more than twenty round flights in Catalinas or Liberators between Western Australia and Ceylon or Karachi since the Qantas service began. This route can never be a joyride. But its bold and skilful and increasingly fast operation as our major external line and the preparations being made to expand it represent probably Australia's most expert and imaginative achievement in civil aviation.

At the beginning of 1945 Captain Crowther returned to Sydney to take up a new position as Assistant Operations Manager of QEA, leaving Captain Ambrose in charge of the Western Operations Division.

The following year when the post-war Australian government announced the creation of a nationalised domestic airline, Qantas Operations Manager and Assistant General Manager Captain Lester Brain took a senior management position with the new Trans-Australia Airlines. Crowther was promoted to Qantas Operations Manager and saw the company through the post-war expansion on new international routes and the introduction of Lockheed Constellations.

Last Two Liberators Arrive

The further two promised Liberators had been much delayed. The third G-AGTI was delivered to Colombo during the first week of December 1945. From there QEA Captain JL Grey continued to Perth on 4-5 December carrying ten civilian passengers. The aircraft continued to Sydney two days later for a pre-service inspection. BOAC assured QEA that the fourth Liberator would be ready early the following year.

When Western Operations Division manager Captain Ambrose received advice in February 1946 that the fourth Liberator G-AGTJ was completing its civil conversion, he was unwilling to wait for British authorities to arrange its delivery. Instead, he gained approval to take a minimum crew to Britain to ferry it to Australia. Like other BOAC Liberators it had been given an overhaul and passenger conversion by Scottish Aviation at Prestwick. When Ambrose reached London, he had a frustrating wait with no information forthcoming on the status of his new Liberator. His personal enquiries revealed it had been completed and test flown by Scottish Aviation but required some final work by a contractor at Croydon Airport, London, before handover to BOAC. Scottish Aviation's test pilots would not take responsibility for landing the Liberator on Croydon's short pre-war runways. Ambrose gained permission to travel to Prestwick airport near Glasgow and flew the aircraft himself to Croydon, where he landed without incident in the deep gloom of a winter afternoon.

Two weeks later in March 1946, Ambrose took G-AGTJ for its acceptance test-flight from

Croydon over northern France but was forced to make a hasty return to the BOAC base at Hurn with his crew suffering food poisoning, blamed on sandwiches eaten before the flight. When all were recovered several days later, Ambrose departed Hurn on the delivery flight to Australia. Qantas senior navigation officer Jim Cowan was rostered to this ferry flight. The first leg was direct to Cairo. They encountered rough weather over France and the Mediterranean and one engine began giving trouble. A landing in the middle of the night in unfamiliar territory was not an inviting option, so Ambrose shut down the faulty engine and feathered the propeller allowing them to continue to Cairo without further problems. From Cairo, they planned to fly non-stop to Ceylon, however the RAF operations office at Cairo insisted the aircraft must land for fuel at Bahrein enroute - obviously unaware of Qantas' long-range capabilities!

G-AGTJ made just one Indian Ocean crossing, Negombo to Learmonth during the delivery flight to Sydney, which Ambrose reached on 25 March 1946. By then Singapore was reopening to airlines and this fourth Liberator joined its Perth sisters on Qantas' new Liberator service Sydney-Darwin-Singapore and return.

The third Qantas Liberator, G-AGTI, seen at Mascot in December 1945. (National Library of Australia)

Chapter 5 Lancastrians: Express Air Service

As the war situation improved in Europe and the Pacific, BOAC and Qantas Empire Airways began planning the resumption of their joint civil airline service between London and Sydney. The Qantas managing director, Hudson Fysh, went to London again in August 1944 for talks with BOAC management and the Air Ministry on joint operation of a new fast service to Australia. It was agreed that Avro Lancastrians would be a suitable stopgap aircraft while awaiting the production of the Avro Tudor airliner, upon which both airlines originally placed their hopes for postwar long-range services. The Lancastrians were transport modifications of Avro Lancaster bombers and could be made available by early the following year.

A telegram from the Director-General of Civil Aviation, London, to the Australian government on 26 December 1944 advised that a number of Lancastrians had been allotted to BOAC for long distance routes, including a through service from England to Australia. The Lancastrians were to be operated by BOAC from London to Karachi, where Qantas crews would take over for the rest of the flight to Sydney. Routing was via Ceylon for the Indian Ocean crossing to Learmonth, then non-stop to Sydney. On 5 December 1944 BOAC despatched Douglas Dakota G-AGHH from Hurn along the proposed Lancastrian route to drop off Lancastrian support parts. It reached Colombo on 13 December.

Although the Lancastrian operation would use the facilities and airfields of Qantas Western Operations Division, the two operations were separately administered within Qantas as the Lancastrian service was controlled from the head office in Sydney. Some Catalina and Liberator crews were initially withdrawn from Western Operations Division to fly the Lancastrians, and eventually all flight crews were transferred to this service.

The Lancastrians

The Lancastrians flown by QEA and BOAC were a much more advanced design than simply a Lancaster bomber with turrets removed and fitted with a new nose. Avro's refinements made it a high-speed passenger transport with an impressive range. It was the first British commercial aeroplane with the capacity to cross the Atlantic to South America. At that time there was nothing else comparable.

The Lancastrian concept began in Canada in 1942 when Victory Aircraft at Toronto modified a Lancaster bomber to a transport. Gun turrets were removed, the nose was faired over and windows were cut into the rear fuselage. After trials by Trans-Canada Airlines, it was flown to England for further modification by AV Roe and Company including

The lines of the Lancaster bomber are clearly evident in this 1945 view of BOAC Lancastrian G-AGMD. (Avro)

The unusual sideways seating layout in the narrow fuselage of a QEA Lancastrian.

the installation of ten passenger seats. Civil registered as CF-CMS, this aircraft commenced a Canadian government trans-Atlantic air service between Montreal and Prestwick, Scotland, from July 1943. Operated by Trans-Canada Airlines, it broke time records on the route.

The success of this stop-gap transport resulted in seven more Canadian-built Lancasters being converted to transports, including a more elegant version with a nose baggage compartment in a redesigned metal semi-monocoque extended nose.

Following an initial British Air Ministry order for 23 Lancaster transports, AV Roe and Company at Manchester put the type into production during 1944, using Lancaster airframes from the end of the wartime bomber production lines. Designated the Avro 691 Lancastrian, the British model had the former bomb bay area modified to a spacious freight bay for mail and cargo alongside two 504-gallon auxiliary fuel tanks which brought the total fuel capacity up to 3,162 imperial gallons. Powered by four 1,635 horsepower Rolls-Royce Merlin T.24 engines, it had an impressive range of 4,150 miles. A total of 82 Lancastrians were built by Avro for the RAF, BOAC and other British airlines.

All armour and armament was removed and the interior of the fuselage aft of the cockpit was stripped out. A passenger section was built in the rear fuselage immediately behind the wing spar. Windows were provided down the starboard side only and the cabin was lined with covered layers of sound-proofing glass, silk, fabric or wool. Because of the narrow fuselage, the passenger compartment contained a row of nine upholstered seats with their backs along the port wall. The seats faced the windows. The seats converted into three bunks and a further three sleeping bunks hinged down from the ceiling. A toilet and wash basin were installed at the rear of the passenger cabin. Passenger luggage stowage was provided in two compartments: one in the elongated nose and the other behind the toilet area in the rear fuselage. This gave each passenger a generous luggage allowance of 50 pounds.

The Air Ministry diverted 21 Lancastrians of its first order directly to BOAC as an interim airliner type to commence fast schedules to Australia. They were registered in the range G-AGLS to G-AGMM to owner British Overseas Airways Corporation, Airways House, Victoria, London. They were to be operated by BOAC from London to Karachi, then by Qantas between Karachi and Sydney. At first, five were allocated specifically to Qantas, however, all Lancastrians were circulated along the full length of the Sydney-London route and were routinely serviced in Sydney by Qantas.

The first Avro-built Lancastrian VB873/G-AGLF was delivered to the BOAC Development Flight at Hurn in February 1945 and two months later BOAC Captain RG Buck flew it on a record-breaking proving flight Hurn-Sydney-Auckland. Departing Hurn on 23 April 1945, the Lancastrian landed at RNZAF Whenuapai, Auckland, in a remarkable three days and twelve hours, with a flying time of 53 hours and thirteen minutes. It set a new standard for Commonwealth services and was a morale boosting exercise for all. The Lancastrian returned via Sydney on 4 May, then routed via Perth, Learmonth, Colombo, Karachi and Lydda to arrive back at Hurn on 11 May 1945.

Despite this record-breaking flight to Auckland,

the proposed Lancastrian London-Sydney service was never intended to continue to New Zealand. Tasman Empire Airways Limited (TEAL) operated a well-established Short Empire flying boat service across the Tasman Sea between Auckland and Sydney. TEAL services between NZ and Sydney were to continue with Sandringhams and Solents until 1954 when replaced by Douglas DC-6s.

Because the Commonwealth airline routes were still under RAF Transport Command control, the Air Ministry also issued these first Lancastrians RAF serial numbers. They were handed over to BOAC as new aircraft in overall natural metal finish, initially painted with an RAF serial, roundel (blue and white South East Asia style) and radio callsign. The military callsigns comprised four letters, for example G-AGLS was OKZS and G-AGLV was OKZV. The first letter indicated RAF Transport Command; the second indicated the type of aircraft ("K" being Lancastrian); the third letter "Z" indicated an RAF aircraft being operated by BOAC and the fourth letter was the individual aircraft identifier, in most cases being the last letter of the civil registration.

These markings were short-lived. By May 1945, a change in Air Ministry requirements resulted in the Lancastrians having military markings removed and replaced by the British civil registration in large lettering on each side of the fuselage and upper wings, with each underlined by a red, white and blue British nationality stripe. A discreet BOAC Speedbird motif was painted in small size on each side of the nose ahead of the cockpit.

VJ Day on 15 August 1945 brought the Second World War to an end, following which RAF Transport Command began relinquishing route control to the civil authorities. The wartime nationality tri-colour stripe underlining the registrations was removed from Lancastrians and Liberators, effective from 13 November 1945. The London-Sydney air service was officially de-militarised effective on 1 February 1946 and later that year BOAC painted fleet names on their remaining Lancastrians.

Avro Lancastrians delivered to BOAC for the shared BOAC-QEA London-Sydney service during 1945-1946 were (as below):

Registration	C/n	RAF serial	Radio Callsign	CofA issued	BOAC name	Fate
G-AGLS +	1173	VD238	OKZS	9.3.45	Nelson	Broken up for scrap at Hurn 1.51
G-AGLT +	1174	VD241	OKZT	20.3.45	Newcastle	Broken up for scrap at Hurn 1.51
G-AGLU	1175	VD253	OKZU	29.3.45		Not flown by QEA: BOAC training aircraft, crashed Hurn 15.8.46
G-AGLV	1176	VF163	OKZV	13.4.45		Sold to Skyways Ltd 5.46
G-AGLW +	1177	VF164	OKZW	26.4.45	Northampton	Broken up for scrap at Hurn 1.51
G-AGLX	1178	VG165	OKZX	14.5.45		Lost without trace Ceylon-Cocos Indian Ocean 24.3.46
G-AGLY +	1179	VF166	OKZY	29.5.45	Norfolk	Broken up for scrap at Hurn 1.51
G-AGLZ +	1180	VF167	OKZZ	2.6.45	Nottingham	To QEA as VH-EAU 11.47 Broken up for scrap at Sydney 9.52
G-AGMA	1181	VF152	OKZA	11.6.45	Newport	Broken up for scrap at Hurn 1.51
G-AGMB	1182	VF153	OKZB	15.6.45	Norwich	Crashed landing Singapore 27.8.48
G-AGMC	1183	VF154	OKZC	21.6.45		Crashed landing Sydney 2.5.46
G-AGMD	1184	VF155	OKZD	29.6.45	Nairn	To QEA as VH-EAS 7.47 Crashed Dubbo NSW 7.4.49 during crew training flight
G-AGME	1185	VF156	OKZQ	3.7.45	Newhaven	Broken up for scrap at Hurn 1.51
G-AGMF	1186	VF160	OKZF	25.7.45		Crashed Broglie, France 20.8.46
G-AGMG	1187	VF161	OKZG	21.8.45	Nicosia	Broken up for scrap at Hurn 1.51
G-AGMH	1188	VF162	OKZH	28.8.45		Crashed Karachi 17.5.46
G-AGMJ	1189	VF145	OKZJ	11.9.45	Naseby	Broken up for scrap at Hurn 1.51
G-AGMK	1190	VF146	OKZK	20.9.45	Newbury	Broken up for scrap at Hurn 1.51
G-AGML	1191	VF147	OKZL	26.9.45	Nicobar	To QEA as VH-EAT 9.47 Broken up for scrap at Sydney 9.52
G-AGMM	1192	VF148	OKZM	3.10.45	Nepal	Crashed Tripoli, Libya 7.11.49

Note: The five aircraft marked "+" were those initially allocated specifically to Qantas.

Lancastrian G-AGLS after its arrival at Sydney's Mascot airport on 17 April 1945. Within a short time, the military markings were removed. (Qantas)

Qantas Prepares for the Lancastrians

QEA Senior Captain Russell Tapp was sent to England early in 1945 to prepare for the new Lancastrians and bring back the first one allocated to Qantas. Following his training at Avro, BOAC and Rolls-Royce, his stay dragged out because of production delays. The enterprising Tapp made good use of that time, arranging to participate in a series of night flying exercises with an RAF Lancaster squadron.

Captain Tapp finally got away for Sydney in G-AGLS/VD238/OKZS, departing Hurn on 11 April 1945. His crew included Captain SK Howard, Radio Officer WR Clarke and two Qantas engineers, Williams and Dusting, who had attended airframe and engine training courses in England. Routing was via Lydda (Palestine), Karachi, Colombo and Learmonth. They reached Sydney during the afternoon of Tuesday 17 April, in an impressive flying time of 58 hours, 12 minutes. G-AGLS was then used for crew training flights and ground engineering training at Mascot.

In Australia, initial Lancastrian training for Qantas pilots had commenced in January 1945, conducted by No. 1 Operational Training Unit at RAAF East Sale, Victoria. The aircraft used was Avro Lancaster A66-1, "Q" *Queenie* (previously RAF ED930), which had been flown to Australia during May-June 1943 for a series of war loan tours. Qantas appointed veteran Captain Orm Denny, who was approaching retirement, to be in charge of the Lancastrian training program. The Civilian Airline Training Flight was established within No. 1 OTU and Denny devised a ground school and flying curriculum. The RAAF record card for A66-1 shows it was allocated to No. 1 OTU on 13 January 1945 and was received at East Sale four days later by the No. 1 OTU Civilian Airline Training Flight.

Among the first to fly the Lancaster at East Sale was Captain Rusell Tapp, who was to be sent to England to deliver QEA's first Lancastrian. Captain Bert Hussey commenced his East Sale flying training on 26 February 1945 and Captain Bert Yates began on 1 March, logging his hours by the Lancaster's radio callsign VM-ZLL. On 10 April 1945 Captain Eric Sims, Captain Jack Bird and Second Officer Ashley Gay carried out Lancaster training at East Sale under instructor Orm Denny. Other Qantas pilots to have initial Lancastrian training at East Sale included Captains SK Howard, OFY Thomas, HG Mills, RJ Ritchie and DF MacMaster.

Three different types seen at Guildford on 7 May 1945: a Lancastrian, Lancaster A66-1 and Liberator G-AGKU. The Lancastrian and Lancaster were being used for Qantas crew training. (Qantas)

During this training period, G-AGLS made some long-distance flights to simulate airline sectors. During the night of 6-7 May, Lancaster A66-1 and the Lancastrian both flew from East Sale to Perth as a training exercise. The Lancaster departed first but the Lancastrian landed at Perth well ahead after a ten-hour flight. Captain Denny was on board, along with DCA officers AR McComb and Hillgedorf and Frank Penny of the Shell Oil Company. The following day the two aircraft departed for Sydney.

Further long-range Lancastrian crew training flights were conducted from both Sydney and Hurn in preparation for scheduled services. G-AGLV departed Hurn on 15 May for a training flight to Karachi, arriving back at Hurn on 20 May. G-AGLT departed Sydney on 26 May for a training flight to Karachi via Learmonth, Ratmalana and Minneriya. Its return flight to Sydney leaving Karachi on 1 June was operated in parallel with the inaugural 5Q1 service, to provide a back-up for the much publicised first service, with both reaching Sydney on 4 June 1945.

In July 1945 Lancaster A66-1 was allocated to No. 7 OTU at Tocumwal for RAAF pilot training in preparation for Australian Avro Lincoln bomber production. However, it was still made available to Qantas. Squadron Leader John Miles was among the RAAF Lancaster instructors. He had been seconded to the Department of Aircraft Production works at Fishermans Bend, Melbourne, for Avro Lincoln production and later recalled:

> Also with the Lancaster I had the job of flying with a number if Qantas pilots giving them experience on the Lancaster prior to Qantas operating the Lancastrian, which was a converted Lancaster. One of the Qantas pilots was a very old friend of mine Rex Nicholl who I knew well from pre-war flying in New Guinea. Rex believed I had been killed in an aircraft accident near Townsville and had helped identify the body, which was apparently mutilated in the face. Rex had not heard of me nor I heard of Rex for a long time when he reported at Fishermans Bend to do his conversion course on the Lancaster. I was waiting for him in the cockpit. He looked at me and damn nearly fainted, he could not believe his eyes. Another of these Qantas pilots was Torchy Uren, an ex-RAAF Beaufighter pilot who I knew very well.

Captain Ken Jackson was a Lockheed 14 pilot with WR Carpenter Airlines on their Sydney-Rabaul route, before it was suspended due to the Japanese invasion of New Guinea in January 1942. He then flew their remaining Lockheed on military charters carrying supplies to Darwin and New Guinea. Ken Jackson remembered:

> When Qantas took us over in 1944, we were amalgamated into their organisation. I continued flying the surviving Lockheed 14 VH-ADT, training new pilots who were starting to be recruited from the air force. In early 1945 the Lancastrian programme was starting, and I was appointed as an instructor on them. I did the training of most of the Qantas pilots down at the air force base at East Sale.
>
> The Lancastrians were a very strong aeroplane, and we were never worried about them breaking up or anything like that. We had a few engine failures, but they were very good aeroplanes.

Captain Dick Mant, also formerly with WR Carpenter Airlines, recalled:

> In June 1945 I went from Qantas Catalinas to Lancastrians after an RAAF conversion course at East Sale. The Lancastrians were a very nice aeroplane to fly, pretty fast, yet very docile particularly in bad weather. A problem was their brakes were very weak. They were air brakes and we often had trouble trying to pull up before we got to the end of the runway. We were operating them Sydney-Learmonth-Colombo-Karachi where BOAC crews would take over and fly it to London. We would pick up the Lancastrian they had flown out to Karachi and fly it back to Sydney.

London-Sydney Lancastrian Service Begins

The inaugural scheduled Lancastrian England-Australia service 5Q1 departed BOAC's wartime English terminus, Hurn airport, Bournemouth, on 31 May 1945. The aircraft was G-AGLV under the command of BOAC Captain E Palmer. At Karachi the aircraft was handed over to Qantas who took it on to Sydney via Minneriya and Learmonth. The pilots for the last sectors to Sydney were QEA Captains SK Howard and HB Hussey who landed at Mascot in the afternoon of Monday 4 June. Only two passengers were carried from England on this first service, Royal Navy officers Admiral CS Standford and Commander H Duncan. They described the flight as "pleasant and uneventful" to waiting reporters. The press also reported that they brought copies of the previous Thursday's *The Times* newspaper, some for delivery to the Duke of Gloucester who had recently commenced a term as Australia's Governor-General.

A fine view of a Lancastrian at Mascot as a Beaufighter passes overhead. Note the BOAC Speedbird motif on the nose. (John Hopton Collection)

The first Qantas westbound service 6Q1 was operated by G-AGLS, which departed Sydney on 2 June 1945 at 0931 flown by Captains OFY Thomas and ER Nicholl and a crew of six. Only two passengers were booked on this inaugural outbound flight. At Colombo Captain KG Jackson and a fresh crew took over for the leg to Karachi, where BOAC crews took over before reaching Hurn on 5 June 1945 at 0758.

The frequency of the Lancastrian service commenced at once weekly in each direction, increasing to twice weekly from 15 July of that year when revised schedules reduced the journey time further, to 63 hours from Sydney to London. The Lancastrian service was at first named the "Antipodes Air Express" before it was later renamed the "Express Air Service". As well as carrying a few high-priority passengers, it significantly increased the transit speed of airmail between Australia and Britain. With such limited seating, the new service was not available to the general public.

With an economic cruising airspeed of 230 knots and fewer refuelling stops, the Lancastrians cut flight times over the 12,000-mile route. This reducing the Sydney to London trip to 63 hours, as compared to the nine and a half days of the Empire flying boats in 1939.

Among new aircrew recruited by Qantas for the Lancastrians was Australian Flight Lieutenant Arnold Easton, DFC, who had completed a tour as a Lancaster navigator with No. 467 Squadron, RAAF, based at Waddington, Lincolnshire. Having risen to the position of squadron navigation officer, he was a particularly valuable addition. In May 1945 in England while waiting for demobilisation, Easton joined Qantas and began Lancastrian familiarisation on 21 May 1945 under QEA Captain RJ Ritchie. It was a test flight from Croydon airport, London, in the next Lancastrian to be handed over to QEA, which Easton recorded in his logbook as VF164 (G-AGLW/OKZW). The aircraft had been at Croydon for modifications before they delivered it to BOAC's main base at Hurn the following day. Over the next few days, he flew with Ritchie's crew on three fuel consumption test flights in the same aircraft from Hurn, the longest being six hours.

On 27 May 1945, Captain Ritchie departed Hurn for Australia in G-AGLW, with Arnold Easton as second navigator. The other crew were Squadron Leader Ratford (co-pilot), J Lawton (navigator) and R Jackson (radio operator). Route stops were Lydda (Palestine), Karachi, Ratmalana, Minneriya and Learmonth before arriving in Sydney on 1 June 1945. From November that year Easton commenced regular flying on Qantas Lancastrians as a navigation officer.

At Karachi, the Lancastrians used RAF Mauripur, four miles from the centre of the city. This large airfield was constructed in 1942 to ease demand on the pre-war RAF Drigh Road with its huge airship hangar. Today Mauripur is Pakistan Air Force's Masroor Air Base. A QEA Lancastrian crew member described a typical arrival at RAF Mauripur from Ceylon:

> After leaving behind the Gulf of Cambay, you could sense the hot, dry air rising off the Sindh Desert. The environs of Karachi materialised out of a dusty reddish haze. With the runways of Mauripur in sight, you could see to the east the great shape of the airship hangar at Drigh Road, a memorial to the ill-fated R-101. Ground transport was always available, you climbed into the two-man carriage behind the turbaned, mahogany-skinned driver - with a flick of his whip he stirred his patient little Indian pony into action and off you trotted into town.

Lancastrian Experiences

The row of nine side-by-side passenger seats facing the starboard cabin wall was unpleasant and claustrophobic for some passengers. On night flights passengers were firmly strapped into the six bunk beds to prevent being thrown out in turbulence. A small steward's galley was installed at the end of the passenger cabin; it was fitted with hot and cold water and a refrigerator. The Lancastrians were generally popular with their pilots, being described as a pilot's aeroplane, with plenty of power yet pleasant to fly.

A shortcoming was ineffective brakes, which faded quickly as they overheated on landing and were to result in a number of runway incidents. Fuel leaks causing a strong fuel odour in the cabin were common. The fuel smell problem was particularly noticeable in turbulence and when fuel was transferred by electric pump from the bomb bay auxiliary fuel tank to the wing tanks as the flight progressed. A "no smoking" rule for the entire flight was strictly enforced.

There were plenty of incidents. During the night of 21-22 December 1945 G-AGLW operating service 5Q72 Karachi-Colombo suffered a serious fire in the starboard outer Merlin, requiring an emergency diversion to Bombay. It landed safely at RAF Santa Cruz in the early morning. G-AGLY was despatched from Karachi that same day as a relief fight, collecting the passengers and mail at Bombay and reaching Sydney on Christmas Eve.

RAAF test pilot John Miles travelled to Karachi by Qantas Lancastrian in May 1946 on his way to England to attend an Empire Test Pilot course:

> The Lancastrian was a most uncomfortable aircraft to fly in as a passenger. Going across India in turbulence, we were strapped in bunks lying down fore and aft, not allowed to move. The bunks were two deep in a very confined space.

Qantas radio officer Dick Labrum also wrote of his experiences:

> The Lancastrian crew comprised two pilots, a navigator, a radio officer and a steward. The captains were mostly veteran Qantas pilots who had logged much of their command time on DH.86s, Empire flying boats and Catalinas. Many first officers and other aircrew had come straight off wartime bomber operations and there was a fair sprinkling of DSOs, DFCs and DFMs among them. A maximum of nine passengers could be accommodated, seated shoulder to shoulder on settees along the port side of the narrow cabin. On night sectors, there was sleeping accommodation for six. There were windows only along the starboard side. A tiny toilet was at the rear of the cabin, just aft of the main door. At the other end of the cabin

Radio Officer Richard Labrum with Lancastrian G-AGML at Mascot. Labrum later wrote an account of his experiences of five Indian ocean crossings on Lancastrians. (Richard Labrum)

was a small galley. Our maximum payload was 3,597 pounds.

The flight deck had no soundproofing, and the noise level was formidable – everything resonated to the rhythm of the Merlin engines. The only way you could converse with your fellow crew members was by intercom. Most of the flight you kept your headphones on, which helped to muffle the noise. The radio officer sat on the port side, jammed between the main spar and a bulkhead on which were mounted the main Marconi Type T1154 transmitter and Marconi Type R1155 receiver, identical to those used in Lancaster bombers. Below these was a small table with the morse key. On the left was a backup Bendix TA-2J transmitter and the winch for the trailing aerial. Immediately overhead was the manually rotated loop antenna to pick up ground radio beacons. To the right of this was the astrodome into which you could comfortably get your head and shoulders - the all-round view from there on a clear day was magnificent. On the other side of the bulkhead sat the navigator and ahead of him the pilots. At night all you could see out two small windows were the inboard exhaust stubs glowing orange with blue centres.

The atmosphere inside the passenger cabin ranged from the chemical smell of the toilet Jeyes fluid aft to petrol vapour forward. When you hit a patch of rough air, the petrol smell became quite disturbing. Smoking anywhere on the Lancastrian was strictly prohibited. There was a heating system of sorts, but it never seemed to thaw out my feet. The aircraft was unpressurised which meant wearing oxygen masks above 10,000 feet.

I flew five Indian Ocean crossings before the traditional route via Darwin and Singapore reopened in April 1945. My logbook tells me that the flights between Colombo and Perth averaged 15.5 hours. I think everybody heaved a sigh of relief when operations were resumed through Darwin and Singapore. It was an easier safer route, but as a replacement hazard we had the notorious Bay of Bengal weather to reckon with.

RAF Indian Ocean Courier Flights

Towards the end of the war RAF Transport Command was being equipped with newer long-range transport aircraft. The command's attention turned to an India-Sydney regular courier leg as an extension of its Trunk Route from Britain to India. The task was given to No. 232 Squadron based at RAF Palam, Delhi, operating Liberator C.VIIs fitted with fifteen passenger seats. These were soon to be supplemented by C-54 Skymasters and the latest model Liberator C.IXs with a single tail fin and a lengthened fuselage.

Flight trials for the new route took place during March 1945. The first was flown by Liberator C.VII EW630 commanded by Flight Lieutenant Saunders, departing Ceylon (RAF China Bay) on 8 March and reaching Perth the following day after a refuelling stop at Learmonth. He left Perth on 15 March on the return flight. Over the next weeks there were two more trial flights:

- Ceylon (RAF Minneriya)-Perth-Sydney, return. C.VIII EW633. Wing Commander Barlow.
- Ceylon (RAF Ratmalana)-Perth, return. Flight Lieutenant Diegnan, RAAF.

The Indian Ocean crossings were considered to be at the limit of the Liberator C.VII range when operated to standard procedures. An RAF airfield was under construction in the Cocos Islands, so the

Liberators of No. 356 Squadron, RAF, at the newly completed Cocos Islands airfield in 1945. (Imperial War Museum)

scheduled courier service did not commence until June when the Cocos airfield was completed.

British military planning had included a forward air base in the Indian Ocean to assist an Allied bombing offensive against Southeast Asian Japanese installations. The Cocos Islands was selected as the best location, no doubt influenced by PG Taylor's survey report. A British Army contingent had been landed on Cocos by January 1945 to start surveying and preparations. During March 1945 a Royal Engineers shipping convoy left India bound for Cocos carrying airfield construction equipment and a force of British and Indian army, navy and air force personnel to construct the airfield on West Island. Work commenced by clearing thousands of coconut palms to construct a single 7,500 feet long runway with a Perforated Steel Planking matting surface. Two large hard-standing areas for aircraft were prepared, as was a control tower building, refuelling facilities and personnel living quarters. When the construction was nearing completion, the Allied Air Commander South East Asia, Air Marshal Sir Keith Park, arrived from India by RAF Sunderland on 29 June 1945 to inspect the new airfield. He also had to finalise terms with the ruling Clunies-Ross family, which included a compensation payment for each palm tree felled.

The airfield was officially named RAF Staging Post Cocos Islands. The first operational sortie from Cocos was made on 3 July 1945 when No. 684 Squadron Mosquito PR Mk.34s commenced a series of photo-reconnaissance missions over Japanese-held areas of Sumatra, Malaya and Singapore. In August Liberators of Nos. 99 and 356 Squadrons took up residence. RAF Spitfires and Catalinas also operated from Cocos. Following VJ Day, British operations were reduced, and the airfield was decommissioned in 1946.

In June 1945 the RAF India-Sydney courier service commenced. The initial route was RAF Palam (Delhi); RAF Ratmalana (Colombo); RAF Staging Post Cocos Islands; RAAF Learmonth; Perth (Guildford); Gawler; Sydney (Mascot), averaging three return flights per month. The Cocos refuelling stop reduced the risks of a longer ocean crossing, but unfortunately on 30 August 1945 Liberator C.VII EW622 crashed on take-off at Cocos with the loss of all on board.

The Cocos Islands was also used occasionally by other RAF aircraft. No. 232 Squadron transport Liberator C.IX JT979 refuelled at Cocos on 15 August 1945 while routing Calcutta-Madras-Ceylon-Cocos Islands-Perth-Sydney-Auckland. Hearing the news of the Japanese surrender, the crew and passengers took two days unauthorised leave in Perth to celebrate the end of the war. They continued to Melbourne (RAAF Laverton) and Sydney (Mascot) where the aircraft received maintenance from resident RAF units. On 1 September while landing after dark in a rainstorm at Auckland (RNZAF Whenuapai), the Liberator ran off the end of the runway and was wrecked, fortunately without injuries.

In December 1945 RAF Flight Lieutenant RP East was flying No. 232 Squadron Liberator C.VIIs at Palam. While operating the India-Sydney courier, his recollection of an encounter with a Qantas Lancastrian highlights the Lancastrians' performance:

> The legs between Colombo, Cocos and Perth could be anything from eight to ten hours of flight time, depending on the weather. The pilots rotated watch every two hours, sitting there monitoring instruments and listening to the constant sound of the engines. The merging of the vast horizon with the Indian Ocean gave a distinct feeling of loneliness out there above thousands of square miles of water.
>
> During one of my watches, halfway to Cocos I was checking some instrument over the far side of the cockpit when something caught my eye to the right of our aircraft. There, about two wingspans away, was an Avro Lancastrian slowly overtaking us. The markings were quite distinctive – a British civil registration and BOAC insignia blazoned on the fuselage side. I tried calling on VHF, but they were not on our frequency. As they pulled ahead, there was a very gentle waggling of the wings, and they slowly became a speck in the distance.[1]

The RAF India-Australia courier service was to be discontinued in April 1946. Liberator VII EW619 operated the final scheduled run on 3 April when No. 232 Squadron Flying Officer Horne departed Sydney for Perth and Ceylon. Meanwhile RAF Liberator transports were active in Australia operating a Sydney-San Diego courier and taking part in post-war demobilisation.

---

1 After the war Ron East migrated to Australia and enjoyed a long civil aviation career as an instructor and DCA examiner of airmen.

Chapter 6 Peace Comes to the Kangaroo Service

Japan's surrender on 15 August 1945 brought the six years of Second World War hostilities to a close. That evening Qantas Liberator G-AGKU was crossing the Indian Ocean as service 3Q97 Ceylon to Perth non-stop. A return to peacetime life was slow for the Australian public because food and fuel rationing was maintained and transport priority was given to repatriating military personnel including the large build-up of British, American and Netherlands forces in Australia.

With the Lancastrians routing Ceylon-Learmonth-Sydney, Qantas decided to drop Perth as its Australian terminus for the twice-weekly Liberator service. The separate company organisation in Perth had been a wartime necessity for the Catalina service that was no longer required. Starting on 30 November 1945, Liberators routed Sydney-Learmonth-Ceylon. The inaugural service 4Q128 was flown by G-AGKU, which left Mascot aerodrome, Sydney, that day under the command of Captains HT Howse and PJR Shields bound for Learmonth and Colombo. Under the service numbering system, it should in fact have been service 10Q1, a point remedied with the next Sydney departure three days later when G-AGKT operated 10Q2 to Learmonth and Ceylon.

Now that Lancastrians and Liberators were crossing the Australian continent from the west coast to Sydney, over 3,000 miles, an optional stop was introduced at RAAF Gawler, 30 miles north of Adelaide. When adverse upper winds reduced the aircraft's range, Gawler was a welcome enroute refuelling stop. There are reports that South Australian airmail was dropped off and collected here but this seems unlikely. The barren Gawler airfield was chosen because of inadequate runway lengths at Adelaide's civil Parafield aerodrome or nearby RAAF Mallala.

The Gawler military airfield had been built in late 1942 on flat land a few miles west of the town. Two hard surface runways suitable for heavy bombers were constructed. It was designated as an RAAF Advanced Operational Base, used by temporary detachments of various RAAF units. There was a control tower and wireless communications hut, but no hangars and few buildings of any kind, although aircraft refuelling facilities were efficient. Personnel from RAAF units operating at Gawler during the war lived in tents at the nearby Gawler racecourse where they used the public toilet blocks with cold showers.

No. 300 Wing, RAF, (later renamed No. 300 Group) in Sydney, formed to provide air transport support for the British Pacific Fleet's planned

Lancastrian G-AGMC during a refuelling stop at Gawler, South Australia, in 1945. (Australian War Memorial)

advance on Japan, discussed use of Gawler with RAAF headquarters in April 1945. A party of 80 RAF ground personnel, known as Staging Post 191, was detached to Gawler from No. 238 Squadron, RAF, then based at Parafield. RAF Liberator transports began using Gawler from June 1945 for refuelling and night stops on RAF courier operations, particularly the RAF courier for military personnel from India to Sydney via Ceylon and Perth. There were plans to build permanent accommodation at Gawler, including comfortable accommodation for aircrew and passengers, but the improving Pacific War situation resulted in accommodation remaining as tents only. RAF Staging Post 191 personnel handled Gawler aircraft parking, refuelling and a crash fire tender, as well as guard duties.

The first Liberator eastbound service 9Q1 from Ceylon to Sydney departed a day late on 3 December 1945 due to G-AGKU requiring a nose wheel change at RAF Negombo. Flown by Captains J Furze and JA Moxham, they arrived at Sydney airport the next day, via Learmonth.

A Qantas ticket dated 13 January 1946 for the Colombo to Learmonth flight. Shortly after this the Indian Ocean flights began a stopover at Cocos Island and used Perth rather than Learmonth. (Qantas)

The introduction of Lancastrians in June 1945 brought the scheduled Indian Ocean crossings to four weekly: one Lancastrian, one Catalina, and two Liberator services, until the Catalinas were withdrawn the following month.

Route Changes - Qantas uses Cocos Islands

As post-war RAF operations at the Cocos Islands reduced, approval was sought to allow Qantas to routinely refuel there for the obvious safety advantages. For Qantas, there was an additional bonus - by refuelling midway across the ocean, the need to maintain the airline's facilities at Learmonth was eliminated. The difficulty and expense in maintaining the remote Learmonth facilities was an increasing burden for the airline and company ground staff resisted Learmonth postings.

From January 1946 the Cocos Islands was an added stop for QEA Liberators and Lancastrians. The RAF Staging Post personnel maintained the airfield facilities for the RAF courier service and occasional military flights. Locating the islands was no longer a significant navigational feat because Qantas aircraft could use the installed radio navigation aids.

The first Qantas service to use the Cocos Islands airfield was 9Q12 Liberator G-AGKT on 14 January 1946, flying Negombo-Cocos-Learmonth-Sydney. It then became a stop for all services in each direction. This new mid-ocean refuelling point triggered a series of route changes. Learmonth was no longer necessary and would be replaced by Perth again. The RAAF encampment at Learmonth kept the airfield operational and it was a reassuring diversion for aircraft on the Indian Ocean crossing.

From 2 February 1946 the new route for Liberators and Lancastrians was Ceylon-Cocos-Perth-Sydney. By this time Qantas flights were rarely using the optional stop at RAAF Gawler near Adelaide, so it was also dropped.

At Perth Airport (the former RAAF Guildford), Australian National Airways was again contracted to provide ground handling and catering for the Qantas flights, which were now just transiting, rather than being based there as before with the

A Catalina amphibian of the No. 112 Air Sea Rescue Flight Learmonth detachment. Aircraft from this unit unsuccessfully searched for missing Lancastrian G-AGLX. (David Vincent Collection)

Liberators. QEA kept a maintenance detachment at Perth Airport. This was just a final temporary arrangement while the airline waited for Singapore to reopen to civil operations, when it could drop the high-risk Indian Ocean route altogether.

Lancastrian Lost

Only two weeks prior to the cessation of the Indian Ocean service and a return to the pre-war route Sydney-Darwin-Singapore-Karachi, tragedy struck: a Qantas operated Lancastrian disappeared without trace over the Indian Ocean.

During the night of 23-24 March 1946, Lancastrian G-AGLX was operating the Ceylon to Cocos sector of service 5Q111. The aircraft's commander was Captain OFY "Frank" Thomas, a widely known Qantas veteran. His crew comprised First Officer NM McClelland, Navigator AA Nuske, Radio Officer WD McBean and Flight Steward R Porteous. There had been a two-hour delay at Negombo, Ceylon, while a radio fault was rectified, but otherwise it was a normal service with five passengers. Among the five passengers was Jack Dobson from AV Roe and Company, England, who was travelling to Melbourne in connection with Australian production of the Avro Lincoln bomber for the RAAF.

Radio contact was lost after G-AGLX made a routine radio report 690 miles northwest of Cocos. When it did not arrive at Cocos, it was posted missing. A huge ocean search was undertaken by a total of thirteen military and civil aircraft. A total of 437 search hours were flown without finding a trace of the Lancastrian. Search aircraft operating from Cocos were four RAF Liberators, one RAAF Catalina and Qantas Lancastrians. From Sumatra the RAF flew a Catalina and a Sea Otter amphibian. The West Australian coastline was searched by an RAAF Liberator and Catalina, as well as MacRobertson-Miller Aviation's DH.86 VH-USW hired for the search. Additionally, all QEA and MMA scheduled flights operating across Western Australia were allocated search areas, with Qantas departures from Sydney being re-scheduled to ensure they arrived over the ocean search areas in daylight.

A report from No. 112 Air Sea Rescue Flight, Darwin, to RAAF headquarters in late March 1946 included the following:

> Catalina A24-112 has been detached to Learmonth since July 1945 and A24-353 was on detachment to Crawley Bay, Perth. Air Sea Rescue activity increased greatly during March. There were two scrambles. One to provide ASR cover to a Douglas aircraft with an engine failure, but it was able to reach Koepang safely. The second was to participate in cooperation with RAF and NEI aircraft to carry out a search for a Lancastrian between Colombo and WA. Two of this unit's aircraft operated under AORWA instructions from Crawley and

Learmonth. Another from Cocos Island. At the time of making this report full details have not come to hand.

Captain Lewis Ambrose flew one of the searching Lancastrians. Several days earlier when he had passed through Ceylon on the delivery flight of Liberator G-AGTJ to Sydney, he spent off-duty time with his old colleague Frank Thomas. Thomas was waiting to take the fateful next Lancastrian service to Perth. When Ambrose reached Mascot in G-AGTJ on 25 March and learnt that Thomas' aircraft was missing he was determined to join the search. Ambrose left Sydney in Lancastrian G-AGLT three days later for the search area. Among the volunteer observers on board was Lancastrian First Officer Alan Wharton, the wartime commanding officer of No. 466 Squadron, RAAF, in England on Lancasters, who was now sharing a flat in Sydney with Frank Thomas.

Among Qantas Lancastrian crews it was assumed the disappearance was due to an explosion in flight. The Lancastrians were notorious for fuel leaks from the fuel line plumbing in the fuselage. The passenger cabin often had the unpleasant strong smell of petrol in flight. One captain recalls an occasion when fuel dripping from a pipe in the cabin was collected by the steward in a saucepan and tipped down the sink in his galley. The cockpit heaters were bled off the engine exhausts, and on occasions, blew sparks through vents in the cockpit. Smoking was strictly forbidden for all on board. A popular theory was that someone may have been smoking a forbidden cigarette in the toilet and ignited the toilet pan fluid, which was later found to be highly flammable.

No trace of G-AGLX was found and no reason was established for its loss. This tragic reminder of the isolation of the Indian Ocean route served to accelerate the changeover to the Karachi-Singapore-Darwin route, which was largely over land or islands with numerous diversion airfields available in an emergency.

Singapore Reopens

The Japanese unconditional surrender on 15 August 1945 meant that occupying Japanese forces in Singapore and throughout Southeast Asia ceased to be combatants. The transition to a peacetime lifestyle was a long and difficult experience. Qantas management was eager to return to the pre-war Empire route via Darwin, across the islands of the Netherlands East Indies to Singapore, then through to Karachi. The RAF had control of all Singapore airfields and in the first few months following the end of the war, military air transport and POW repatriation was its only priority.

In early August, Captain Russell Tapp, who had been appointed Western Operations Division manager, was transferred to the Ceylon manager position. The prospect of resuming Qantas services through Singapore instead of the hazardous Indian Ocean route was foremost on his mind:

> I had only been in Ceylon a few days when VJ Day was celebrated and Singapore was liberated. I managed to get myself on to an RAF Sunderland and got to Singapore. When I arrived, I was able to get in touch with people I knew who arranged transport for me from the Seletar flying boat base. I went to the Seaview Hotel, where we used to stay and found the old manager but there were no rooms. I went back to the RAF whom we used to depend on for everything, saw the senior operations officer, Group Captain Gordon Pirie and he invited me to stay in the mess. He then asked "Have you got a uniform?". I had my Qantas uniform in my case, and he told me to put it on. I had my First World War campaign ribbons and, seeing them, Gordon said "Put them on quickly, nobody else has them!"

> The AOC was Air Commodore the Earl of Bandon, a fine man who was always known as "Bandy", who introduced me to Colonel Webster who had been a solicitor in Singapore before the war. He found me an old Chevrolet car that had been used by the Japanese. The starter had been removed by the Japanese, so I

had to twiddle two wires together to start it, but there were few other people there with any cars at all. I sent a cable to Sydney telling Qantas I was in Singapore to start the flying boat service again and was given approval to stay. The Royal Navy had its vessels completely covering the harbour and there was nowhere to land our flying boats. I arranged a high-level conference with the navy and brought along the Singapore harbourmaster, a retired RN captain with a stutter, who the navy thought would be on their side. He wasn't, he was on my side. He told them to "move their b-b-b-bloody ships".

Thanks to arrangements made personally by Tapp in Singapore, Qantas was approved to bring a delegation from Australia by flying boat for talks aimed at resuming airline services. On 8 October 1945 Short Empire VH-ABG *Coriolanus* alighted at RAF Seletar flying boat base, after a five-day flight from Sydney via Darwin, Morotai and Labuan. The crew comprised commander Captain KG Calwell, Captain FA Reeve, Radio Officer GW Mumford, Radio Officer JF Christie and Flight Engineer WE Wilcox. Company personnel carried were Captain WH Crowther and Traffic Superintendent G Allen together with Ground Engineers AK Stone, ND Capel, A Koh, AP Lopes, WW Skyes and RS Carr. Also on board were DCA Airline Supervisor Mr R Christie and Shinty Colvill, the former QEA agent at Batavia. Captain Crowther, now QEA Operations Manager and Superintendent Empire Operations Division was in charge of the delegation. The ageing flying boat was used because war-ravaged runways were in poor condition at Singapore and enroute through the Netherlands East Indies.

Bill Crowther wrote of his impressions of Singapore in *Empire Airways* magazine:

> To fly over it, the island looks exactly the same as when we left it in February 1942, with the exception of the Kallang and Changi airstrips, but when walking through the city one quickly notices that property has been allowed to deteriorate badly and the predatory Japanese have removed such things as metal lamp posts, elevators and other essentials.

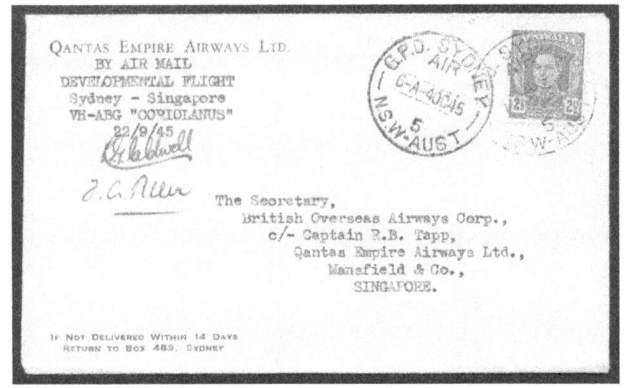

An official Qantas letter addressed to Captain RB Tapp in Singapore as carried by the flying boat Coriolanus during its October 1945 flight. It is signed by pilots KG Caldwell and FA Reeve.

The party made inspections of the airfields and facilities at Singapore, and the Qantas members commenced arrangements to secure office and accommodation space in the city. Mr Christie made arrangements for point-to-point radio communication between Singapore and Darwin and the establishment of aviation weather broadcasting facilities. Negotiations commenced with this visit resulted in agreements for Qantas to use both land and water aerodromes early the following year. *Coriolanus* returned to Sydney on 17 October, bringing home some Australian Prisoners of War.

Captain Tapp stayed on in Singapore as the first postwar QEA Singapore Manager. With 18,000 hours logged he was content to cease flying duties. He remained at Singapore until October 1948 when he was recalled to Sydney to establish new routes to Manila, Hong Kong and Japan with Douglas DC-4s.

Singapore's pre-war Kallang aerodrome, close to the city, was unusable. Allied prisoners-of-war who had been forced to maintain Kallang's runway under brutal conditions had clandestinely sabotaged the construction beneath the surface. One of the early Allied aircraft arrivals at the end of war, an RAF Liberator bomber, was damaged on landing when its main wheels broke through the runway surface. Initial post-war civil airline services would have to use Changi aerodrome while Kallang's runways were rebuilt and the stripped terminal building repaired.

Lancastrian G-AGMD on the perforated steel plating runway at RAF Changi circa 1946. (Geoff Goodall Collection)

Changi aerodrome, which the Japanese had built during the war using Allied prisoners of war from the nearby Changi prison as slave labour, was being made operational as an RAF station. Agreement was reached that the facilities of RAF Changi would be made available for civil airlines as an interim until Kallang became available a year later.

BOAC commenced post-war flights to Singapore with Sunderland/Hythe flying boats operating schedules Poole (UK)-Singapore return. The first service was 13F1 flown by G-AGKV *Huntingdon* departing Poole on 31 January 1946 and reaching Singapore on 5 February. Two days later it departed Singapore as 14F1 for Poole. From May that year the Hythes continued on to Sydney to make a flying boat through service between Britain and Sydney.

The first post-war Qantas landplane through Singapore was Lancastrian G-AGMH, which landed at Changi on 22 February 1946 while conducting a survey flight over the "new" route to Australia. Under the command of Qantas Captain BC Sims, the survey flight returned via Darwin on 26 February, before reaching Sydney the following day.

From April 1946 all Lancastrian services were routed Karachi-Singapore-Darwin-Sydney, instead of the Indian Ocean crossings. The first Qantas scheduled Lancastrian service through Singapore was service 7Q1 operated by G-AGME on 9 April 1946. It arrived with a cracked windscreen, which was replaced at Changi. The first westbound Qantas service was 8Q1 flown by G-AGMC, which landed at Changi on 10 April 1946 from Sydney and continued on to Karachi.

The first Liberator Sydney to Singapore service was 12Q1 on 7 April 1946, operated by G-AGTI under the command of Captain JAR "Allan" Furze. It reached Singapore on schedule the following day, compared with three days for the pre-war flying boat service. For the first few weeks the 12Q service continued from Singapore to Negombo, Ceylon, where it turned around as 11Q Negombo-Singapore-Sydney. It is believed the Negombo extension was primarily to facilitate the removal of QEA staff, records and equipment as operations through Ceylon were discontinued. The first Liberator eastbound service 11Q1 was flown by G-AGKU, which departed Negombo on 9 April for Singapore and Sydney. Flying time was 10 hours 41 minutes Singapore-Darwin, then 9 hours 33 minutes to Sydney, arriving the following day.

At that time Darwin was in a bad state with evidence of war's destruction everywhere, including funnels and masts of sunken ships visible across the harbour. There was little accommodation in town and QEA crews were at first billeted at the Berrimah hostel, the former 119th Army General Hospital.

From April 1946 QEA was scheduling three Lancastrian and two Liberator services through

Singapore each week. From May the Liberators terminated at Singapore while the Lancastrians continued on to Karachi where they were handed over to BOAC crews. During this period, Qantas made Singapore the crew rest location between Karachi and Sydney. Initially the Singapore QEA transit accommodation was the Seaview Hotel, a pre-war grand establishment. But the Japanese had departed only eight months before, and it was now a stripped shell of its former self, an austere transit billet for crew and passengers. Navigator Allan Hughes recalled that period:

> The Seaview Hotel was still in a hell of a mess from the war. We used to take food, crockery, cutlery, soap and other things off the Hythe flying boat services and the stewards would get local food. In this manner we kept the passengers, and ourselves, happy in this huge somewhat derelict hotel.

Qantas crews' memories of this Changi period included Japanese POWs being used to clean their aircraft's passenger cabins, while other POWs were out on the runway, swinging sledgehammers to belt out the dents in the perforated steel plating runway matting after each aircraft landed. Also remembered was the Changi crew transport to take them to and from their hotels, which was a dilapidated truck with wooden slat seats on its open tray, leaving them exposed to the elements while lurching along unmaintained roads.

The Lancastrian "Express Air Service" to England had been a successful postwar stopgap, providing a speedy mail and priority passenger service which helped a return to peacetime air services between the two countries. The Lancastrians were far from economical, with the annual loss of £1,400,000 on the Kangaroo Service being accepted by British and Australian governments as the price for maintaining a fast prestige service to Australia. From 31 May 1946 the London terminal for the Lancastrian service was changed from Hurn Airport, Bournemouth, to the newly constructed London Airport at Heathrow.

Until September 1946 only government-approved priority passengers had been allowed to travel on the Kangaroo Service. This wartime restriction was now dropped and passage made available to the public, but at an airfare for only one class of travel, first class. Qantas advertisements in newspapers and travel agents' windows invited bookings with "no more delays waiting for travel permits". Unrestricted international air travel had resumed.

Even with the air route reverting to Darwin and Singapore, Qantas continued to operate the world's (then) longest regular non-stop airline sector: the Lancastrians flew direct from Singapore to Karachi, 2,975 nautical miles in an average of fourteen flying hours. A long 1,635 nautical mile leg over water between Port Swettenham (Malaya) and Yanam (India) was often flown in monsoonal weather conditions. At take-off on this sector a fuel load of 3,000 imperial gallons was carried. Not infrequently the Lancastrians stopped at Madras enroute for fuel. The same engine operation plan as used on the Catalinas and Liberators was adopted: the speed set at the beginning of the flight was carefully maintained throughout; engine power was continually reduced as the fuel load decreased to maintain an indicated air speed of 210mph, as compared with speeds of up to 260mph on other sectors of the route.

There were Lancastrian accidents. On 2 May 1946 G-AGMC was damaged at Sydney while returning from a pilot training sortie at RAAF Williamtown. To avoid delays caused by other aircraft at Sydney Airport, Qantas regularly conducted circuit training at Williamtown during air force stand-down periods. Civil aerodromes at Narromine and Dubbo were also used. On the landing approach to Sydney G-AGMC was allowed to undershoot and the starboard undercarriage struck an earth bank just short of the runway and collapsed. The aircraft skidded off the runway with the starboard wing and power plants badly damaged. The four pilots on board were uninjured, but the aircraft was declared a write off. On 31 January 1947 G-AGMB swung during landing at Sydney, causing an undercarriage collapse. It was repaired and returned to service before being written off on 27 August 1948 due to a runway overrun when landing at Singapore on a freight flight. These events received surprisingly little newspaper coverage.

Meanwhile Qantas senior management was making major decisions on the airline's postwar development and choices of new aircraft types. Production delays and disappointing performance of the Avro Tudor II resulted in the first major Qantas break from British allegiance – cancelling the Tudor order and selecting instead the American Lockheed L.749 Constellation. This caused a political rift between the two governments, quelled only when BOAC itself also ordered Constellations.

While awaiting delivery of the new Constellations for the Kangaroo Service to London, Qantas and BOAC agreed to an interim combined flying boat service between Australia and England, using BOAC Short Sunderlands upgraded to civilian "Hythe" standard airliners. These were operated similarly to the pre-war Empire flying boat service, with Qantas operating Sydney to Singapore, and BOAC between Singapore and Poole, England. The Hythes retained their British civil registrations.

Despite seating for only nine passengers, the Lancastrians continued a fast landplane service, mainly aimed at speedy airmail. The Karachi handover between QEA and BOAC Lancastrian crews was later dropped, and Qantas crews flew the entire route Sydney-Darwin-Singapore-Karachi-Lydda-Malta-London. Lancastrian time for the straight-through journey was scheduled as 63 hours, with flying time averaging 48 hours.

Final Indian Ocean Service

The reopening of Singapore for civilian airliners in April 1946 enabled termination of the Indian Ocean route. The official date of cessation of the Indian Ocean Service from Ceylon to Australia was 8 April 1946. The name "Kangaroo Service" transferred to the new route to England from Sydney via Darwin and Singapore.

The final Indian Ocean Service was operated by Liberator G-AGKU, which flew 10Q37 from Perth to Cocos and Ceylon over 5-6 April 1946, in a total time of 18 hours 47 minutes.

A vital era of Qantas' proud history had ended. What were its achievements?

- Established a separate airline operation, in wartime secrecy, nearly 2,000 miles from its main base, at a time when company resources were stretched tightly. The Perth engineering organisation maintained the aircraft to the strictest standards, often with limited access to parts.
- Reopened air communications between Australia and England via the much shorter Middle East route, assisting the Allied war effort.
- Regular operation of the world's longest non-stop airline sector, with the added risk of an ever-present enemy. Devising an engine operation plan which helped achieve this long-distance flying operation, extending the Catalina's range beyond previously accepted limits.
- Reinforced the operating partnership between BOAC and Qantas Empire Airways, which had been severed by the fall of Singapore. Without this wartime collaboration, QEA's hold on the post-war Australia-England route may not have been so entrenched.
- The professionalism of magnificent aircrews, often facing duty periods of over 30 hours for each flight (or over 20 hours in a Liberator), flown in virtual radio silence.
- Aerial navigation excellence with few ground navigation aids, applying highest standards of precise astronavigation to allow a confident regular flight schedule.
- Accomplishing 525 Catalina and Liberator crossings of the Indian Ocean without injury to passengers or crew. While the tragic loss of Captain Thomas' Lancastrian must be considered, by then the Lancastrian service was not under the control of Western Operations Division.

The final and lasting achievement of the Indian Ocean service was that it established Qantas' solid reputation for reliability and safety, which enabled the airline's postwar expansion to a worldwide route network. The aviation reputation of today's Qantas Airways long-distance international services was built on the foundations of the dependability pioneered across the Indian Ocean during the Second World War.

Qantas closed their facilities at Perth, Ceylon and Learmonth as Western Operations Division finalised its operations during April 1946. For almost three years it had successfully plugged the wartime air route gap between England and Australia. Division staff returned to Sydney, where Captain Ambrose was appointed Qantas Senior Flight Captain. The professionalism of the division was recognised by keeping the key personnel separate from the Lancastrian service – to regroup at Qantas Sydney Head Office to form the high-level project team planning the pending introduction of Lockheed Constellations.

The Liberators and Lancastrians were now based at Sydney Airport. The Liberators operated a twice-weekly service to Singapore via Darwin, while the Lancastrians flew a thrice-weekly service via Darwin to London. More than six years were to pass before Perth was again to see Qantas services, when the new "Wallaby" service to South Africa commenced with Constellations from 1 September 1952.

Effort Unrecognised Officially

No person involved in the unique Qantas Indian Ocean operation received any kind of official government recognition for extraordinary service. Despite various government awards recognising Australian civilians' bravery in war zones, inexplicably none were granted to the Qantas aircrew who flew through enemy patrolled airspace, nor engineering ground staff who worked tirelessly on vital maintenance and overhaul. The best offered by the Department of Air was Returned Serviceman status to selected Indian Ocean personnel including the appropriate campaign medals for the different theatres of war in which they were engaged. These were awarded merely for being there, not for any specific service above and beyond the call of duty.

Qantas aircrews wore standard RAAF issue uniforms, with their QEA ranks and insignia: although they were civilians, they had been signed up as RAAF reservists. Had any been captured they would not have been legally entitled to status as prisoners of war. On most flights they flew military personnel and military supplies. The Catalinas retained their RAF roundels and serial numbers, designating them as military aircraft. As well as the long arduous hours of work necessary to complete an ocean crossing, a distinct type of bravery was essential for the men flying the route.

Qantas Managing Director Hudson Fysh, frustrated by his inability to gain official recognition for the men who made the Indian Ocean service such a vital part of the war effort, wrote a personal letter to selected men dated 30 April 1947:

Dear ……

Since May 1945 I have been engaged in an effort to obtain recognition by way of decoration or official commendation for certain Qantas Empire Airways staff who performed meritorious work during the war on active operations, and I have just received advice which seems finally to preclude this, though I pushed the matter hard, both in Australia and in the United Kingdom.

Many members of QEA richly deserve official recognition, the absence of which I believe is a serious omission. As a possible source of satisfaction, I am informing by letter all members who were so recommended by me, though many others also deserve recognition.

You were recommended for your work in respect of the valuable Indian Ocean service, which was a great wartime contribution.

(Signed) Hudson Fysh

Fysh had earlier ensured they at least received recognition from their own airline. He created a special award, in the shape of a star, called the Long Range Operations Gold Star which became popularly known as the "Crowther Cross". It was issued from December 1943 to all aircrew to have completed four return trips on the Indian Ocean service, or three RAAF Catalina delivery flights from Honolulu, subject to the following conditions:

1. The gold star is to be worn only when in uniform and is to be worn immediately over the wings or other insignia worn, or if the left breast is bare, over the left pocket.

A QEA pilot's wings and the Long Range Operations Star known as the "Crowther Cross".

2. Once a crew member has qualified for the star, he is always entitled to wear it when wearing QEA uniform even if engaged on different duties.

The skilled navigation employed on the Indian Ocean services was recognised by the aircrews' peers in April 1949 when Captains Crowther, Tapp and Ambrose were jointly awarded the "Johnson Memorial Trophy". This premier award for aerial navigation was created in memory of the navigator of the British airships R100 and R101, Squadron Leader EL Johnson. This was the first award of the trophy since the Second World War. The citation stated:

> The Guild of Air Pilots and Air Navigators of the British Empire considers that the work performed by Captains WH Crowther, RB Tapp and LR Ambrose constitutes an outstanding feat of Airline Navigation, and we have great pleasure in adding their names to the names of those distinguished navigators who have preceded them in establishing similar performances in the past.

No recognition was given to the two meteorological men whose weather forecasting made the Indian Ocean operation possible. RAAF Squadron Leader John "Doc" Hogan was the senior forecaster in Perth, and Wing Commander Arthur Grimes was his opposite number with the RAF in Colombo. A strong bond of respect and friendship grew between QEA crews and these two men at each end of the Indian Ocean route. The service depended largely upon the accuracy of their forecasts of winds and weather, and the secret reports from the observers on the Cocos Islands. Their forecasts and advice were trusted implicitly. Captains Crowther, Ambrose and Tapp described the work done by Grimes and Hogan as "wonderful".

When Ambrose was posted to England in 1949 to become QEA London Manager, he took the opportunity to approach the Royal Meteorological Society to attempt to gain recognition for these two forecasters. After addressing a meeting of the society, praising their work and describing the weather aspects of the Indian Ocean operation, Ambrose awaited a decision. To his surprise and embarrassment, sometime later he learnt that he had been awarded a fellowship of the society, while nothing had been afforded to Grimes or Hogan!

Thirty years after the Indian Ocean service ended, former Western Operations Division personnel had their first official reunion. It was a formal dinner organised by Qantas' General Manager, Captain RJ "Bert" Ritchie, himself an Indian Ocean Catalina veteran. The allocation of funds for the reunion was one of his last acts before retiring. On 28 June 1976, at Qantas' head office, Qantas House, 70 Hunter Street, Sydney, the men gathered to renew friendships and revive memories. Many travelled from distant parts of Australia while Captain Tapp made the journey from his retirement home in England. The event was a great success.

Six Qantas Indian Ocean service veterans at a reunion dinner at Qantas House in 1976. From left to right are RJ Ritchie, WH Crowther, G Mumford, RB Tapp, F Furniss and R Senior.

PBY-6A VH-EAX painted to represent G-AGIE Antares Star at the Qantas Founders Museum at Longreach, Queensland, in 2020. (DL Prossor)

Forty years after the Indian Ocean service, a memorial plaque was unveiled at the site of the Qantas Nedlands Marine Base on the Swan River, Perth. On 29 June 1984 a small ceremony was held to unveil the plaque mounted on a large stone, attended by Captain Bill Crowther, his wife Meeta and Qantas Deputy Chief Executive Officer Ron Yates. A Qantas Boeing 747 departing Perth Airport flew overhead and a radio message from the captain was broadcast to those at the ceremony.

At the same time Qantas also arranged for a Qantas Catalina Memorial at the city of Galle, Sri Lanka, close to Koggala Lake. Both plaques include the same words:

> Commencing on 29 June 1943 Qantas Empire Airways operated the World's longest regular non-stop service Perth to Ceylon (Sri Lanka) a distance of 3,513 miles (5,632 kilometres). The initial services were operated by Catalina flying boats, known as the flight of the Double Sunrise. These missions were flown in complete radio silence and without any radio navigation aids. The story of the Indian Ocean services is one of triumph over adversity and the highest standards of aviation endeavour.

The very successful Qantas Founders Museum at Longreach, Queensland, where the airline had its origins, was determined to acquire a Catalina. Years of negotiations came to fruition in 2009 when the museum purchased a retired fire bomber PBY-6A Catalina which had been retired in the open weather in Spain for twelve years. After an eventful delivery flight with a long delay in Thailand because of engine faults, the Catalina (EC-EVK, registered VH-EAX for the ferry flight) arrived at Longreach on 14 September 2011. It has since had side blisters and a nose turret installed and is painted to represent G-AGIE *Antares Star*.

Statistics

- Distance Ceylon-Perth: 3,513 nautical miles (5,632 kilometres)
- Distance Karachi-Ceylon-Perth: 4,970 nautical miles (8,000 kilometres)
- Total Catalina Indian Ocean crossings: 271
- Catalina miles flown: 1,380,119
- Longest flying time: 31 hours 45 minutes
- Shortest flying time: 22 hours 46 minutes
- Catalina passengers: 524
- Catalina freight: 14,383 pounds
- Catalina mail: 113,760 pounds
- Total Indian Ocean crossings 1943-1946 (Catalina/Liberator/Lancastrian): 824

The Qantas Liberators made 274 crossings when based in Perth as 3Q/4Q services and a further 74 from Sydney as 9Q/10Q. Liberators carried a total of 1,010 passengers across the Indian Ocean. Lancastrian crossings totalled 232, including the ill-fated 5Q111 G-AGLX.

Qantas documents quote a grand total of 4,412,015 route miles being flown to carry 5,412 passengers. In the First British Commonwealth and Empire Lecture, titled "Australia in Empire Air Transport", delivered by Hudson Fysh in London on 13 November 1945, he provided data applicable up to 31 March 1945: 235 Catalina trips had been scheduled, of which 223 were completed. Figures for the Liberators were 115 and 103 respectively.

Despite the longer flight time required to complete a Catalina crossing, the Liberators accumulated the greater number of Indian Ocean flying hours. As at 2 May 1945 Catalina G-AGIE *Antares Star* had logged 2,490 hours (of which only 200 hours were not over the Indian Ocean).

Qantas Liberator VH-EAJ at Mascot circa 1947. It served for a few years as a freighter before it was broken up in 1950. (John Hopton Collection)

In contrast Liberator G-AGKT by 4 June 1946 had logged 3,818 hours (including 268 hours flown in RAF service prior to it being received by QEA). G-AGKU by 15 June 1946 had logged 3,392 hours (including 72 hours with the RAF).

At the time, the Catalina route from Perth to Ceylon was the world's longest regular airline sector. It is still the longest non-stop time in the air for airline operations, a record which will probably never be broken, at least until inter-planetary flights become a regular occurrence.

Liberators and Lancastrians Subsequent Use

When the Indian Ocean route was discontinued in April 1946, the Lancastrians flew a thrice-weekly Sydney to London service via Darwin and Singapore, changing crews at Karachi with BOAC as before. The four Liberators were immediately transferred to a new twice-weekly Sydney-Darwin-Singapore service. This was an interim measure to get the airline back into civilian international services, replaced during the year by the shared BOAC/QEA Short Hythe flying boat service between Sydney and London via Singapore. One Liberator was withdrawn from use in May, one in June and the last two during August 1946.

The Liberators were to be returned to BOAC when retired. Because they had been purchased new by the British Government, there were no Lend-Lease requirements attached to their disposal. The four aircraft were parked at Sydney Airport while waiting for instructions from London. The British CofAs for the oldest two with higher airframe hours, G-AGKT and G-AGKU, expired on 27 June 1946 and 26 August 1946 respectively. This pair was sold to a scrap metal dealer and unceremoniously broken up at Mascot and carted away. The fuselage sections were resold, one became a rabbit trapper's hut at Dungong, New South Wales, and another a fish and chips stand in the Sydney area.

Liberators G-AGTI and G-AGTJ were retired from the Sydney-Singapore passenger service in August 1946 and parked at Sydney Airport with the other two Liberators. Qantas could see a need for these two lower-hour aircraft to assist in post-war expansion. Their purchase was negotiated from the British Air Ministry at a heavily depreciated price. G-AGTI and G-AGTJ were ferried to Brisbane for Australian certification overhauls as freighters at the Archerfield Qantas workshops. First to emerge was G-AGTJ in April 1947 which was now registered VH-EAJ with a metallic finish with Qantas Empire Airways markings. Just prior to the overhaul, G-AGTJ had been used for crew training at RAAF Williamtown on 5 March 1947. After test flights at Archerfield, the Australian CofA as VH-EAJ was issued on 28 April 1947. G-AGTI followed two months later as VH-EAI, receiving its CofA on 26 June 1947.

The two remaining Liberators gave valuable service for crew training and freight work. With the high engine failure rate experienced by the newly introduced Lockheed Constellations, the

Lancastrian VH-EAU at Bofu, Japan, during a courier service in 1948. (Air History.net photo archive)

Liberators were particularly valuable for carrying Constellation spare engines. Qantas and BOAC operated a pool agreement to provide overhauled Wright Duplex Cyclone engines, tyres and other spare parts at ports along the Constellation route to London. The Liberators transported these power plants, each engine weighing over a ton, until both aircraft were finally retired at Sydney during 1950 and broken up. Check and training captain Bob Gray remembered:

> I used to go out to Parkes, NSW, with a Liberator quite a lot, doing licence renewals for the Constellation first officers. I would spend seven or eight hours at Parkes doing circuits and bumps and three engine approaches and so on. I liked the Liberator, which had four good Pratt & Whitney R-1830 engines, the same engine as the Catalina and DC-3. The Liberators were also used for carrying spare engines for the Constellations for engine failures up the track to London.

During 1947 the Australian government requested Qantas to establish a fast weekly courier service between Sydney and Japan to carry mail and personnel to the Australian contingent of the British Commonwealth Occupation Force. A fast courier was needed to supplement the RAAF courier schedules to Japan operated by C-47 Dakotas with numerous stops along the long route. With newly ordered Lockheed Constellations and Douglas DC-4s dedicated to prestige passenger services, Qantas decided the best aircraft for the job would be the Lancastrian. Flying operations and engineering personnel knew the type well and a strong spares stock was held – there should be little delay in getting them back into Qantas service.

Three Lancastrian Mk.1s from the original Express Air Service partnership with BOAC were purchased from the British Air Ministry. A fourth was a later model Lancastrian Mk.3 G-AHBW purchased from Silver City Airways Ltd at Blackbushe, with whom it was named *City of London*. This last Lancastrian also was no stranger to Australia, Silver City Airways having conducted high-speed long-range flights from England to South Africa and Australia for its mining company financial backers. The four aircraft with their Australian registration dates were:

- VH-EAS 20.10.47 ex G-AGMD handed over to QEA at Hurn 2.9.47
- VH-EAT 20.10.47 ex G-AGML handed over to QEA at Hurn 12.9.47
- VH-EAU 20.10.47 ex G-AGLZ handed over to QEA at Hurn 28.10.47, arrived Sydney 2.11.47
- VH-EAV 2.2.48 ex G-AHBW Silver City Airways, Blackbushe, purchased 21.1.48

Qantas did not give individual aircraft names to these Lancastrians. Qantas engineering replaced the sideways facing lounge chairs with a row of forward-facing single seats on each side of the fuselage, to carry thirteen passengers. A row of matching windows was installed down the port side.

RAAF roundels were painted on the fuselage sides while operating the Japan courier service. The first proving flight to Japan left Sydney on 29 November 1947 in VH-EAS crewed by captains LR Ambrose and JG Morton, with navigator JB Cowan. The inaugural scheduled courier left Sydney on 16 December 1947 crewed by captains JG Morton, ARH Morris and FT Boyce.

The Japan RAAF courier service route was Sydney (Mascot)-Melbourne (Essendon)-Darwin-Manila (Clark Air Force Base)-Bofu RAAF base, Japan. The destination changed from May 1948 to the larger military base at Iwakuni. Captain John "Minnie" Morton recalled his time on the courier:

> Of all the aeroplane types I flew, the Lancastrian was the one I most enjoyed flying. Whether it was the loud purr of those four Merlins, or because it was so strongly built, or the fact that it is not so docile as the DC-3, DC-4 or Super Constellation, I don't know. Going through cumulonimbus, the Lanc seemed to just shrug its shoulders, saying "Toss anything you like at me, I can take it." We used to stay overnight at the large USAF base Clark Field just out of Manila. The Americans had a great respect and liking for our Lancastrians and on my last flight out of Clark before we moved to Manila International Airport I was asked if I would like to buzz the field. Flying just above that long runway from a dive was the fastest I went in the trusty old Lanc.
>
> On some southbound trips we used to pass relatively close to Okinawa and within the control of the Naha USAF Base, where they had some of the first F-86 Sabre jet fighters. The Sabres had a practice of using us for interception exercises, flying in pairs and when they caught up with us, we would all have little chat on one of the control frequencies. So I would go on to rated power as soon as we entered their airspace to gain a much higher speed than cruising and make them work harder.

The four Qantas-owned Lancastrians were used on a variety of tasks, including pilot training, freight carrying and proving flights to open new routes like Sydney-Norfolk Island. When BOAC withdrew its

Lancastrian VH-EAV at Mascot with its belly pod for carrying Constellation engines. (John Hopton Collection)

last freighter Lancastrians during 1950, Qantas introduced a scheduled Sydney-Singapore Lancastrian cargo service from 6 September 1950 to add capacity to the Constellation mainline services through Singapore. Freight demand was high and only a month later the Lancastrians were replaced by DC-4s from 14 October 1950.

VH-EAV was later modified with a large clamshell pod under the fuselage to carry a spare Constellation engine. Known as the "Pregnant Lanc", it was developed by Qantas for the primary purpose of positioning spare engines along the Constellation route Sydney-London. Engine failures were common in the early years of Constellation service, each Wright R-3350B Duplex-Cyclone 18-cylinder radial engine weighing over a ton. This engine pod modification was designed by Qantas aeronautical engineer Ron Yates. The modified Lancastrian had its first test flight at Sydney on 17 May 1949, flown by Captain John Morton, with Ron Yates on board as flight test observer. Flight tests found that degradation to directional and longitudinal stability was remedied by minor flap extension. The "Pregnant Lanc" entered service on 18 August 1949 carrying a spare Constellation engine to Karachi.

Lancastrians revisited the Indian Ocean during proving flights for Qantas' proposed Perth-South Africa route. Captain Lewis Ambrose took VH-EAS to Johannesburg and return via the Cocos Islands and Mauritius in November 1948, followed by Captain John Morton in VH-EAT in February 1952.

The adventurous lives of these four Qantas Lancastrians were punctuated by a series of runway accidents. The "Pregnant Lanc" VH-EAV had three such events, none of which was attributed to the engine pod. The first was only weeks after commencing service as an engine carrier: on 15 September 1949 when landing at Sydney after a test flight, Captain J Cole was unable to stop by the end of the runway and deliberately ground-looped the aircraft on to soft earth, collapsing an undercarriage assembly. The damage was soon repaired.

On 13 April 1951, after departing Singapore (Kallang) for Darwin, VH-EAV returned with a Merlin shut down. The captain made the approach at a higher-than-normal airspeed to allow for the dead engine. However, he was unable to stop by end of the runway and deliberately ground-looped the aircraft, collapsing the starboard undercarriage on the steel-matting runway. The aircraft was again repaired and returned to service fitted with a new belly pod manufactured by Qantas Engineering.

The career of the "Pregnant Lanc" came to an end in November of that year when it ran off the runway on take-off at Sydney Airport. It was conducting a series of flights to Auckland, New Zealand, each carrying a Bristol Hercules engine from two Tasman Empire Airways (TEAL) Short Solent flying boats stranded at Rose Bay. Each aircraft's four Hercules engines had seized after their oil systems had been topped up from incorrectly labelled drums of an unsuitable liquid. It was a major commercial blow for TEAL, for which the oil supplier accepted financial responsibility.

Just before midnight on 17 November 1951, VH-EAV had reached 60 knots on take-off at Sydney when No. 1 engine lost power and Captain Mullins was unable to stop a swing to port. The aircraft ran off the runway, the landing gear collapsed and it came to rest on its belly across a drain. The crew were not injured but the aircraft was written off. Cause of the power loss was found to be a mechanical failure in the propeller reduction gear. Qantas now used its Douglas DC-4s when engines needed to be transported.

The final fates for the Qantas Lancastrians were:

- VH-EAS crashed Dubbo NSW 7.4.49. Training flight, ground-looped on landing, landing gear collapsed, aircraft destroyed by fire. No injuries.
- VH-EAT retired at Sydney Airport. Broken up for scrap 11.52.
- VH-EAU retired at Sydney Airport. Broken up for scrap 11.52.
- VH-EAV "Pregnant Lanc" ran off runway on take-off at Sydney Airport 17.11.51, landing gear collapsed. No injuries. Write-off and broken up for scrap.

When the last two Lancastrians were scrapped at Sydney, one fuselage was acquired by Gifford Carter who had it cut it into three sections. The tail end went to a mini-golf course at Manly Beach, Sydney, and the other two sections were mounted on trailer bogies as mobile Fish and Chips vans equipped with gas burners. One was reportedly still in service at the Maitland, NSW, showgrounds as late as 1974.

***

The wartime Indian Ocean services were etched in Qantas Empire Airways' corporate memory as it expanded to become today's Qantas Airways. So many exceptional aircrews gained their experience on those long dark ocean crossings. Captain RJ "Bert" Ritchie who later rose through Qantas management to be appointed Chief Executive Officer in 1967, flew 81 Indian Ocean crossings in Catalinas and Liberators.

Qantas founder and long-time Managing Director Sir Hudson Fysh, commenting on the Indian Ocean Service many years later, said:

> I have always felt that this was the most fascinating and romantic undertaking ever performed by Qantas.

QEA Lancastrian VH-EAV after a landing accident at Kallang, Singapore, in April 1951. The aircraft was repaired but was written off seven months later after a take-off accident at Sydney.

Appendix 1

Following is a detailed account of a typical Catalina service from Perth to Karachi and return titled *Order of the Double Sunrise*. Written by Senior First Officer EH "Ted" Neal, it was published in *Shell Aviation News* in 1972 and is reproduced here with permission of the magazine editor and the author:

> An icy cold atmosphere of -9 degrees centigrade permeated the flight deck as dawn broke over the Indian Ocean. The eastbound Qantas Catalina *Altair Star* began a gradual descent from 12,000 feet towards an unbroken layer of stratocumulus cloud which shrouded the Australian coastline on a July morning 33 years ago. It marked the closing stage of a regular but little-known wartime civil transport operation. The date was 18 July 1945. After two years of service, operations over the 3,060 nautical mile non-stop Indian Ocean route from Australia to Ceylon were now to become another chapter in civil aviation history.
>
> The service had been launched by Qantas Empire Airways in July 1943, restoring the Australia-UK air route which had been forced to cease nearly eighteen months previously. Our new one-hop route ran south of the war zone from a base on the Swan River at Perth, Western Australia, to Koggala Lake in Ceylon. Later, at the request of BOAC, the service was extended to Karachi 1,400 miles further on.
>
> Operations were commenced with two Catalinas supplied by BOAC. Ultimately, five Cats were employed on the route. The first four were service aircraft and were partially modified for the job before departure from Great Britain, final modifications being carried out by Qantas in Australia. The last Catalina was modified to Qantas requirements before its delivery flight from the United Kingdom.
>
> To obtain the greatest possible payload, equipment and fittings not required for civil operations were removed. These including the aerofoil de-icing boots, as it was considered unlikely that severe icing conditions would be encountered. Eight auxiliary fuel tanks were fitted in the area under the flight engineer's station, bringing the total fuel capacity to almost 2,000 Imperial gallons.
>
> Three "C" class flying boat passenger chairs were placed athwartships and two stretcher type bunks were fitted in the section forward of the blister compartment. A toilet for passengers and crew was provided aft.
>
> After stripping the aircraft of all surplus weight, the available payload amounted to about 1,100 pounds and take-off weight varied between 35,000 and 35,400 pounds. The payload usually consisted of three VIP passengers (each limited to 30 pounds of baggage), diplomatic mail, urgent freight - mainly for the armed services - and "Airgraph" mail. Initially the Catalina operation was on a weekly basis, later increased to three return services per fortnight and finally to a twice weekly return service. The crew complement was three pilots, navigator, radio operator and flight engineer.
>
> **Flight Planning**
>
> Pre-departure routines for each of these flights followed a similar pattern. Early in the morning the captain and navigator would prepare the flight plan, based on information supplied by the Meteorological Service at Perth, or the RAF in Ceylon. These long-range forecasts, covering such a vast expanse of ocean, were remarkably accurate considering the scarcity of information available. The generally adverse wind component affecting the westbound flights called for careful

flight planning for each of the twelve zones into which the route was divided, as meteorological and navigation conditions varied considerably according to the season.

No alternative landing areas were available along the route. The Cocos Islands could not be used as a diversion because they were located near the midway position of the route and were passed during the hours of darkness. Moreover, radio silence prevailed, as enemy held territories lay only a few hundred miles to the north-east. An additional reason was that the westbound journey was flown at an altitude of between 1,000 and 2,500 feet and both low and high cloud often obscured the stars for long periods, which prevented the accurate astronavigation so necessary when attempting a landfall of this nature.

Embarkation at Perth was by means of a jetty, to the seaward end of which was attached a pivoted pontoon which could be swung alongside the moored aircraft. When preparations for departure were complete the pontoon was swung clear, engines were started and warmed, the endless forward mooring line leading to two dolphins (one on each bow) was cast off, and the stern of the aircraft was released by means of a running bight. All three mooring lines could be handled through a shore-based winch.

When departing Perth in the summer months the aircraft would usually glide across glassy calm, the river reflecting the countryside around, with overhead a clear blue sky. These conditions, though serene, were not conducive to short take-off runs and could lead to prolonged full power operation. Yet an extended take-off involved no problems of distance as there were several miles of water available without any high surrounding obstructions.

Departures six months later would present a different picture. The blue sky would have changed to grey clouds, moving with strong cold west to southwest winds and frequent showers. These winds were the cause of many anxious glances at the temperature gauges while taxiing downwind and crosswind. Weather of this nature could delay a departure, as the previous glassy calm of the river would become a grey green turmoil, which made the heavily laden Catalina appear more like a submarine than a taxiing flying boat.

Take-off time was reckoned from the moment the throttles were advanced to maximum power of 1,200 bhp and a manifold pressure of 48 inches, until the aircraft became airborne. Depending on wind velocity, air density and the take-off technique adopted, this time varied between 50 and 80 seconds. On occasions longer times and higher powers were recorded for Perth-bound departures from the confines of Koggala Lake in Ceylon.

Reliability

Delayed departures were rare and very few services were abandoned after becoming airborne. Most of the delays could be attributed to adverse weather conditions: the few cancellations arose mainly from engine or airframe problems, such as fuel leaks from the integral wing tanks.

This high degree of reliability was achieved despite the fact that, when the Catalinas were introduced, spare parts and experienced aircraft engineers were difficult to obtain in the Pacific region. Furthermore, the main Qantas workshops were more than 2,000 miles away on the east coast of Australia.

The Perth-Ceylon run was actually begun without adequate ground servicing facilities. Its success was thanks mainly to a small but dedicated group of Qantas engineers at Perth, and - at the start of the operation - the willing assistance of the US Navy's flying boat base nearby. We also enjoyed the same level of support from the RAF at Koggala.

Undoubtedly the very low incidence of service delays from mechanical or airframe failure could be attributed to the high standards of ground maintenance, and also to the care taken by the aircrews in setting up the engine power requirements in flight. This care was reflected in reports received from the RAAF maintenance unit at Kalgoorlie, where stripped-down engines were found to be in remarkably good condition. No doubt a further factor was that as these engines were used on long individual flights, the number of full power take-offs was less (for a given total of hours) than those of engines removed from service aircraft.

The original Pratt & Whitney Twin Row Wasp R-1830 engines were of American manufacture. Australian built engines were later installed and gave the same excellent service - certainly a tribute to the young Australian industry.

Navigation

West-bound flights were made at a relatively low level because of adverse upper wind components, but flights in an easterly direction were made at much higher levels to take advantage of the more favourable winds at altitude. The usual procedure on east-bound runs was to set course for Cape Inscription, the northern tip of Dirk Hartog Island, which lies about 410 miles north of Perth.

A rhumb line track was always flown. During daylight, navigation was by dead reckoning, with log and chart entries being made every hour based on observations of drift sights, direction of sea and cloud changes. Drift readings were taken through the bomb aimer's window in the nose of the aircraft. Astro sightings of the sun (when cloud permitted) produced an awkward position line with regard to the aircraft's track and generally were of little value. At night, position plots from three-star astro sightings were made. When necessary, drift sightings were carried out by means of an astro compass, which could be mounted at either blister hatch. A flame float, which burst into light when it hit the water and gave off thick smoke in daylight, was used in conjunction with the astro compass.

Crew members quickly became all-rounders. The three pilots relieved the navigator, radio operator and flight engineer. The captain and the first officer took over the navigator's duties while he rested. The second officer relieved the radio operator and flight engineer and was responsible for providing meals for crew and passengers. The navigator was always on duty for a period after sunset and immediately before sunrise so that the aircraft's position could be accurately plotted at these times. He was also "on deck" at the time of the midway forecast. The radio officer was on duty for some hours after take-off and before landing and for the receipt of the midway forecast.

The flight engineer's main duties arose during the first and last hour of flight, and also during the period in which the auxiliary fuel was transferred to the main tanks. This procedure was begun about five hours after departure.

Broadly speaking, crew members would have a couple of one-hour breaks off duty during daylight. At night, the two breaks were split into one of two hours, and another of one hour.

The flight plan for the eastbound flights was based on making a landfall at Cape Inscription, from which position a slight alteration in course was made for the final run down the Australian coast to Perth. During the winter months the coast was generally obscured by strata cumulus cloud, in consequence of which alterations to course were made on dead reckoning.

The oft-made statement that "Australia couldn't be missed" could be very misleading. Perth lies about 180 miles north of the southern tip of the continent, and, if the oblique angle of approach formed by the direct Ceylon-Perth track is pictured, it will readily be appreciated that a landfall 400 miles north of Perth was a better form of insurance than planning to "go direct".

Typical flight – Westbound

What was it actually like on one of these long, lonely Indian Ocean crossings, without radio, steering by the stars and facing the vagaries of tropical fronts? Perhaps some description of a westbound crossing may provide the answer.

Altair Star was nearing the end of her take-off run. The deep roar of the engines blended with the higher note from the propellers to create an atmosphere of power and expectation. The flying boat escaped from the restraining hold of the water as light back pressure was applied to the control column. The pressure was then eased and a small power reduction made, and the aircraft was allowed to gather speed before a reduction to rated power was made and a gradual climb begun. As a very gentle turn was made onto a compass heading of 315 degrees the city of Perth, beside the tranquil waters of the Swan River, passed under the starboard wingtip.

The Catalina settled down to cruising altitude of 1,500 feet and a true airspeed of 118 knots suspended between the blue sky and the deeper blue of the ocean below, as the coast of Australia faded away on the starboard quarter. The passengers and crew prepared for the long journey ahead.

For the next 25 hours they would be in an isolated world of their own. Even the radio waves had been silenced, thanks to the turmoil of the times. There was a distinct nip in the air at sixteen degrees Celsius as the limited comforts of the uninsulated and unheated aircraft were investigated. A goodly supply of rugs and pillows was unpacked, but the standard air force flying suits and boots would not be required on this journey, the air temperature rising as the aircraft moved into tropic latitudes.

Catering arrangements for the passengers and crew were good - the food hamper and assorted large vacuum flasks contained tea, coffee and soup. Cold fruit drinks were supplied in lieu of soup on the southbound flights from Karachi and Ceylon.

A small electric hot plate was fitted in the blister compartment. But the low temperatures, combined with altitude on the eastbound flights and the draughty conditions of this part of the aircraft made the heating of soups and drinks a protracted business.

Sets of chess and draughts and plenty of magazines were provided for the passengers, to help while away the monotony of the long flight. A thorough briefing was always given on life jackets, handling of the inflatable raft and emergency procedures in the event of forced alighting on the open sea.

Discreet indications were given that the two bunks were, primarily, there for the use of the crew. Occasionally, however, a passenger would slip into one that happened to be unoccupied. No crew member had the heart to toss him out, often preferring to bunk down on mail bags under the navigator's table or on similar cargo stowed in what had been the forward gun turret compartment. This latter position entailed two occupational hazards - if in daylight the navigator wanted to take a forward drift sight the recumbent body was smartly moved aside; and secondly, when the aircraft entered heavy rain the sleeping beauty would generally get a gentle trickle of water down his neck.

The sun was low in the west, the long south-westerly ocean swell slightly blurred by waves, while overhead cumulus clouds, increasing in depth, were becoming webbed together by a tracery of cirrus. Altostratus was also appearing, portending the meeting of the northern and southern hemisphere airs at the Inter-Tropic Front ahead.

A flight time of 24 hours 50 minutes had been forecast, a very fast time for a westbound flight. The first 1,050 miles, except for a very mild front several hundred miles astern, had been flown at a progressively increasing ground speed of from 121 to 132 knots, the true air speed of 118 knots being maintained by reductions of power as fuel was consumed.

A last faint gleam of twilight remained. The temperature had risen to 21 degrees Celsius, yet the increasing cloud allowed only an isolated star or two to show themselves for a few seconds. Optimistically, the navigator watched the blackness above for the chance to obtain an astro-fix. The aircraft's position had been plotted by dead reckoning since leaving Perth 1,145 miles to the southeast, so it was important that an accurate position should be established, as sun shots were of only limited navigational value.

The navigator had a hard task on the westbound flights at low altitude. Long periods were spent watching for a suitable star to appear through breaks in the overcast, and under these conditions the shooting of a three star "fix" could be a long procedure, aggravated by turbulent conditions in which the sextant bubble and star were unwilling to unite.

The passengers dozed fitfully as the aircraft wallowed on through the bumps, the inky blackness of the night being thrown into relief by intermittent lightning which illuminated the heavy tropical rain streaming off the windscreen and momentarily revealed the propellers as pale green discs.

As *Altair Star* approached the midway position, the radio operator decoded the forecast broadcast from Ceylon, giving probable weather for the remainder of the journey and terminal conditions for the estimated time of arrival. About this time a very cryptic message was transmitted, in plain language, giving only the last two letters of the aircraft's registration, its altitude and ground speed. No acknowledgement of any transmissions was made or received.

The navigator, who may have been lucky enough to snatch a couple of hours in the bunk, was woken about one hour before sunrise so that the important pre-dawn position could be astronomically obtained. A good fix was essential for the last 550 nautical miles of the flight. The island of Ceylon is often difficult to sight from any great distance, visibility being poor through the haze and rain under a low cloud base.

During the hours of darkness, the ground speed had decreased from 135 to 122 knots as the wind veered from the southern quarter to the northeast. *Altair Star*, now some four tons lighter and 3,015 miles from Perth, should have sighted Ceylon about 45 miles ahead. Yet the gradually lowering cloud base, together with its attendant shadows and showers, caused several false sightings. At a distance these features can bear a close resemblance to a distant coastline.

For some time, an increased airspeed of 120 knots was maintained though the ground speed had decreased to 113 knots. Even so, an ETA of 0800 hours LST indicated the possibility of a record westbound flight of less than 24 hours.

There was always an air of restrained excitement among the passengers and crew, waiting for the first landfall after the long flight. The engineer from his position in the tower was the first to sight terra firma, a few degrees on the starboard bow, Dondra Head emerged through a camouflage of heavy rain. The automatic pilot was disengaged, magnetos tested, and to clear the scattered cloud that rushed to meet and envelop the aircraft, a descent to about 900 feet was initiated.

Dondra Head lighthouse passed under the starboard wingtip, followed closely by the small town of Matara with its white thatched dwellings, and the white bell-shaped Buddhist temple nestling among the palms. Sweeping on across the waters of Welieama Bay, the aircraft now faced a palm-covered ridge a mile or so ahead, behind which lay Koggala Lake. The level rays of the early morning sun lit the blue sea contrasting with the dark shadows beneath the clouds; the lush green of island scenery was obscured by a mantle of cloud.

Koggala Lake came into view, surrounded by palm trees and dense tropical jungle, with a tree-covered bank at one end holding back its placid waters. The port wing dipped in a graceful turn as our Catalina skirted the lake, past moored RAF Sunderlands and Catalinas reflected in the calm waters. From a jetty at the base a crash launch trailing a V shaped wake sped out to take up position near the alighting area.

The deeper tone of the engines now changed to a higher note as propellers were put into fine pitch, and the whine of an electric motor in the wing above announced the extension of the wingtip floats. Banks and reed-covered shallows flashed close beneath the bow, the control column was eased back, and a soft hissing sound came from under the hull as the throttles were closed. With decreasing speed, the flying boat continued across the lake, the crash launch following astern. Finally, as though tired from her journey, *Altair Star* sank off the step into the following roll of her wake.

The aircraft slowly turned, rocking gently in the waves of her landing run, and taxied towards the crash launch circling the mooring buoy. Streaming two drogues astern, *Altair Star* moved lazily towards it. The second officer, positioned on the sponson, quickly secured the mooring pennant and laid it over the bollard. Mixture controls were moved to idle cutoff, and the rumbling murmurs of the idling engines died away to the last few clicks of the valves. A hush settled over crew and passengers alike.

The silence was broken by a gentle thud as the tender came alongside to take everybody ashore. Before disembarking, the captain presented the passengers with certificates proclaiming that, as they had spent more than 24 hours in the air, they were eligible to become life members of the rare *Secret Order of the Double Sunrise*. Our flight time was 24 hours 3 minutes, and buoy time 24 hours 21 minutes. This was possibly a record westbound flight as generally the time was in the order of 27 hours.

Onwards to Karachi

At Koggala Lake the palms and moored flying boats were reflected by the setting sun as the passengers, mainly military personnel, boarded *Altair Star* for a twelve-hour flight to Karachi. It was to be a coastal flight with favourable winds. Since the Catalina's arrival earlier that day she had been serviced and, though spick and span, the interior was hot and sticky.

The crew for this return flight to Karachi were those who had arrived at Koggala from Perth a few days earlier. The run to Karachi, could be flown by either one of two routes - the first via Minikoi, an island north of the Maldive group, was 1,500 nautical miles; or the second, skirting the west coast of India, a distance of 1,400 nautical miles. The former avoided the late afternoon and evening build-up of massive and very active cumulonimbus cloud along the southern coast of India and seaward from Cape Comorin. On this occasion, as only isolated cumulus cloud had been forecast, *Altair Star* would follow the west coast route.

The aircraft lifted her relatively light load of 29,640 pounds off the water in 39 seconds and climbed to a cruising height of 2,000 feet where the air temperature was 26 degrees Celsius. Course was set to clear the southern tip of India by several miles. After being airborne for about an hour lightning could be seen on the starboard bow, the sky ahead appearing clear in the light of the full moon. With the coastal thunderstorms disappearing astern, course was altered more to the north. This would bring the aircraft within sight of the Indian coast, near Calicut.

As we neared the coast, the many fishing boats with their swaying lanterns resembled a glow-worm grotto. It was the night of the full moon celebrations; each village and town along this thickly populated coastline had a bonfire, some casting their reflection in the nearby sea or inland waterways. Bombay glowed in the distance while the silver moon lit the empty waters across the Gulf of Cambay.

Dawn was breaking as the flying boat approached the delta of the great Indus River, revealing a wilderness of snake-like muddy creeks. Out to sea a steamer and several dhows could be seen approaching Karachi, our destination, as the aircraft made a low wide sweep around the port. Through the haze the giant airship hangar – built for the ill-fated airship R-101 - grew visible, making even the newly arrived Super Fortress bombers look like toys. *Altair Star*'s step was feathering the dirty yellow water as she sped by a long berth of ships to port, including the burnt-out hulk of the White Star liner *Georgic*, which had been dived-bombed and later towed from the Middle East. To starboard lay a dredge, several lighters and a motley collection of small craft, which the control launch was trying to keep clear of the alighting area. The Catalina taxied to her buoy between the moored BOAC flying boats, *Canopus* and *Golden Hind*, impressive company.

Southward, again

Owing to the delayed arrival of the service from the United Kingdom carrying urgent diplomatic mail for Australia, our scheduled departure the next evening was postponed until dawn the following day.

As take-off time approached, the stars shone weakly through an opaque atmosphere. The moorings were slipped, and the aircraft was towed from her confined area to the flare path some distance away, where the tow was dropped and engines started. The last of the flickering orange flares sped by as the aircraft lifted from the water and swung onto course bound for Ceylon. Dawn illuminated the tortuous waterways and salt swamps of the Indus. The delta region southward to the Gulf of Cutch receives no benefit from the rain-bearing winds of the South West Monsoon. The winds blowing on this coast originate in the dry interior north and west of Karachi: their passage over the Arabian Sea is of short duration, and therefore they are dry. This area which forms only a small part of the arid Sindh Desert is virtually treeless, any hills appearing as a mass of wind eroded rocks.

As *Altair Star* cleared the delta swamps, the nearest land was about 30 miles to the northeast, and the next landfall would be made at Dwarka Point. The next 140 miles to Diu Head at the entrance to the Gulf of Cambay would be along a low-lying sandy coast with little of interest except for a mosque at Porbandar, birthplace of the late Mahatma Gandhi, from which flew a yellow flag as large as the building itself.

An hour and a quarter after Diu Head, Bombay lay on the port beam. Course was altered more to the south as the beautiful scenic coastline of southern India ascending to the Western Ghats came into view, the peaks of which rise several thousand feet. The aircraft passed a mile or so westward of Rajpuri Island, behind which lies a large estuary, and ahead the wide Savitri River, scenery typical of the coast southwards to Cape Comorin. In many places the coast has sandy beaches and rocky headlands, with jungle-like vegetation growing right down to the sea indicative of the very heavy rains originating from the South West Monsoon.

Shortly after Savitri River a small rock-walled island fort appeared, rising sheer out of the sea. Several cattle grazed on the lush green grass within the walls. Passing down the Malabar coast, Goa was sighted, at that time a Portuguese colony: then on to Calicut, where Vasco da Gama landed in the 15th century. During one flight in July the previous year, an Annular Eclipse of the sun was witnessed by the crew, the coastal areas taking on a ruddy hue while the distant mountain peaks appeared to be in bright sunlight.

Along most of the coast southwards there is a fairly continuous waterway consisting of lakes and connecting rivers which provide calm protected waters for the transport of merchandise between the towns along this populous coast. By mid-afternoon, with cumulus cloud building from Cape Comorin, *Altair Star* changed course for Koggala Lake and headed across the Gulf of Mannar.

Dusk was falling as the aircraft made a direct approach to the landing area. The small white temple on the lakeside stood out in stark contrast to the darkening forest background. On this occasion we had arrived in the evening, instead of the morning, so the waiting Perth-bound flight would be delayed 24 hours. *Altair Star* would now be serviced and test flown in preparation for her departure for Australia in about two days' time.

Farewell

At 0600 hours local on 17 July 1945, *Altair Star* was swinging lazily to the buoy on Koggala Lake in a light west to southwest breeze. As the passengers and crew went aboard, the sky showed signs of altostratus with, beneath it, some drifting cumulus. Now was the time when, normally, the connecting aircraft from Karachi could be expected. But there was no sound of approaching engines in the northern sky this morning, for *Altair Star* was preparing to depart on the very last of the Qantas "Double Sunrise" services across the breadth of the Indian Ocean.

An unusual number of RAF personnel were watching from the jetty and along the lakeside near the base. As *Altair Star* headed out across the water, her escorting crash launch sounded three long blasts of farewell. Sluggishly the laden aircraft gained speed: the tempo quickened as the tail was raised, and on the step she sped down the lake, past the crash launch towards the fast-approaching palm trees at the end. Sixty-six seconds after power was applied, the 35,000 pound aircraft climbed away in a very gentle turn, skimmed the tops of the trees and set course for Perth at a height of 1,000 feet.

The equator was crossed a little more than four hours later, at which point the aircraft resumed climbing. Breaking through a seven-eighths stratocumulus cloud layer at 5,000 feet we discovered a continuous mantle of altostratus cloud above concealing the morning sun.

Seventeen and a half hours after take-off, the temperature had fallen to freezing point. Our last astro-fix had indicated a ground speed of 145 knots, the true airspeed being 126 knots. The sky was crystal clear, the inhabitants of the heavens now sharply defined in an atmosphere of minus four degrees Celsius. Areas of stratocumulus cloud, which increased as the last 600 miles of the route to Australia were covered, drifted by below.

With the plotting of the pre-dawn fix completed, it was apparent that an early arrival of 0849 hours local would be achieved - a fast flight of 24 hours 19 minutes, but not a record. By now the temperature had dropped to minus nine degrees Celsius. Dawn was beginning to break as the Catalina was prepared for descent over the Indian Ocean. It was appropriate that *Altair Star* should close the operation, since it was she that begun those eventful journeys twenty-four months before.

Perhaps the ghostly sounds of those reliable Catalinas, named after the stars that served them so well, may still be heard over the lonely Indian Ocean.

The Koggala-Perth sector described above was 1Q134 17-18 July 1945 in *Altair Star*. At the time it was to be the final Catalina service. The crew comprised Captain PJR Shields, First Officer EH Neal, Second Officer McLennan, Flight Engineer MacDonald, Radio Operator Balfour and Navigation Officer Keith. However, Qantas scheduled one further Catalina service, 1Q135 operated by *Rigel Star*, which arrived in Perth on 24 July 1945.

Appendix 2

QEA Western Operations Division Personnel Chart

Manager Western Operations
Captain WH Crowther,
then, Captain LR Ambrose

Senior Station Engineer	Senior Flight Engineer	Station Engineer Colombo
Nedlands	NW Roberts	B Jenkins
NW Roberts		then, C Short
Senior Navigation Officer	**Senior Engineer Nedlands**	**Traffic Superintendent Perth**
JLB Cowan	CC Sigley	RG Cochrane
Senior Route Captain	**Senior Engineer Koggala**	then, AJ Quirk
RB Tapp	CW South	**Traffic Superintendent Ceylon**
	then PH Compston	RG Cochrane

QEA Aircrew Indian Ocean Operations 1943-1946

Lancastrian crews during the Indian Ocean routings are included, although the Lancastrians were not operated by Western Operations Division

Captain	First Officer	Second Officer	Navigation Officer	Radio Officer	Engineering Officer
Ambrose LR	Cole JH	Anderson RJ	Bailey JW	Balfour HJ	Bull RD
Bird JA	Corkran TJ	Crowe KM	Bartsch HJ	Balfour FJ	Clark CW
Crowther WH	Couchman FA	Hale KG	Brearley JH	Beresford BPA	Furniss FS
Fairservice RJ	Dunne RA	Hicks PB	Cowan JLB	Clarke WR	MacDonald JD
Furze JAR	Elliot JG	Kiernan JT	Grainger D	Gillman	McCourt TE
Grey JL	Fielding F/Lt	Lee HG	Hoare GA	Hartley J	Malcolm DW
Griffith JF	Furniss FS	MacKenzie R	Hollingsworth F/Lt	Hatton K	Martz W
Koch AA	Gay AL	McLennan CC	Keith JW	Hocking H	Maskiell FJ
Howard SK	Locke HB	Riordon JR	Lawton JH	Hurndell CA	Muir DC
Hoskins GP	Morris ARH	Simpson A	Leek RN	Jackson RT	Nixon F
Howse HT	McClelland NM +		Lyons D	Jackson RJ	Roberts NW
Howson PW	Mather MV		Mitchell FC	Lander RE	Rowe J
Hussey HB	Martin JR		Nuske AA +	Lanyon GW	Sigley CC
Jackson KG	Moreland G		Patterson CJ	Louttit LW	South CW
MacMaster DF	Neal EH		Pearn SN	McBean WD +	Willcox WE
Mant RO	Nettleship GM		Richards EC	McCaskill FC	Wood J
Mills HG	Peirce IC		Rose NS	MacKenzie HS	Woods H
Moxham JA	Radford HW		Sander FE	Mumford GW	
Murray JN	Roberts R		Walker BS	O'Dwyer KA	
Nicholl ER	Senior RC		Walsh ED	Richards EC	
Pollock JC	Shannon DS		Wright GP	Robertson S	
Ritchie RJ	Swift F/Lt			Saxby AM	
Ross HH	Wharton A			Shirley WA	
Shields PJR				Stewart AW	
Solly JW				Stewart JR	
Tapp RB				Tracy RDR	
Thomas OFY +				Turner GW	
Thurstun GR				Weston RMH	
Uren RF				Young R	

+ Crew of missing Lancastrian G-AGLX

Appendix 3

RAAF Catalina delivery flights

Qantas Captain HB "Bert" Hussey's 1941 pilot logbook pages show how QEA aircrews were rotated between the Sydney-Singapore airline service and RAAF Catalina deliveries. Also shown is the Qantas route extension Sydney-Singapore to Karachi. The final entries clearly show the immediate disruptions caused by Japan's entry into World War II: the 7 December 1941 Pearl Harbor attack followed immediately by offensives against Malaya, Burma, Thailand and the Netherlands East Indies along the Qantas airline route.

March 3-6	Short Empire VH-ABF	Singapore-Sydney	Surabaya, Dili, Darwin, Townsville
April 8-12	Catalina A24-3	Honolulu-Sydney	Canton Island, Noumea
April 15	Catalina A24-3	RAAF Rathmines	Crew training
May 2-5	Catalina A24-4	Honolulu-Sydney	Canton Island, Noumea
May 29-June 2	Short Empire G-AETV	Sydney-Singapore	Townsville, Darwin, Surabaya
June 9-12	Short Empire G-AETV	Singapore-Sydney	Surabaya, Darwin, Townsville
June 26-27	Catalina A24-6	San Diego	Test flights
June 28-July 5	Catalina A24-6	San Diego-Sydney	Honolulu, Canton Island, Honolulu (return), Canton Island, Noumea
July 14-18	Short Empire G-AEUB	Sydney-Singapore	Townsville, Darwin, Surabaya
July 21-24	Short Empire G-AFPZ	Singapore-Sydney	Surabaya, Dili, Darwin, Townsville
Aug 13-16	Catalina A24-8	Honolulu-Sydney	Canton Island, Noumea
Sept 3	Catalina A24-10	Honolulu	Test flight
Sept 5-8	Catalina A24-10	Honolulu-Sydney	Canton Island, Noumea
Sept 18-22	Catalina A24-12	Honolulu-Sydney	Canton Island, Noumea
Oct 4-7	Catalina A24-16	Honolulu-Sydney	Canton Island, Noumea
Oct 15-17	Catalina A24-17	Honolulu	Test flights
Oct 18-20	Catalina A24-17	Honolulu-Sydney	Canton Island, Noumea
Oct 24	Short Empire G-ADUV	Sydney	Test flight Rose Bay
Oct 27-Nov 3	Short Empire G-ADUV	Sydney-Karachi	Townsville, Darwin, Surabaya, Singapore, Rangoon, Calcutta
Nov 4-15	Short Empire G-AEUF	Karachi-Sydney	Calcutta, Rangoon, Singapore, Surabaya, Darwin, Townsville (rest Singapore)
Nov 26-30	Short Empire G-AFRA	Sydney-Singapore	Townsville, Darwin, Surabaya
Dec 5-7	Short Empire G-AEUI	Singapore-Karachi	Rangoon, Calcutta
Dec 10-13	Short Empire G-AEUC	Karachi-Batavia	Calcutta, Akyab, Port Blair, Sabang, Singapore
Dec 15	Short Empire G-ADVB	Singapore-Batavia	
Dec 19	Short Empire G-ADHM	Batavia-Surabaya	

Appendix 4

Nedlands Catalinas Scuttled, a report by Arthur W Jones

The following candid first-hand account describes the scuttling of the four Catalinas retired after the war at the Qantas Nedlands base at Perth. At the time of these events the author was Sergeant AW Jones, Fitter-Air Gunner, serving with No. 43 Squadron Catalinas at Rathmines, New South Wales. Checking against his personal logbook, Arthur first recorded this account in 1991 at the request of the Australian Catalina Association.

On 3 January 1946 I was in the Officers' Mess at Rathmines, and the Orderly Officer came in and asked if there were any West Australians in the Mess. My first thoughts were don't volunteer for anything, but, what the heck. I said "Yes, I am a West Aussie and a Fitter Air Gunner." My mate Johnny Evans, who was a Fitter Engineer, was with me and said he was available for whatever was going on if no WA personnel were found. The Orderly Officer told us to report to the Orderly Room next morning. We duly front up and they said "You two can have the job" and so pilots Flight Lieutenant Hodson and Flight Lieutenant Withell, and the two of us were told it was a top-secret job in WA and none of us had been home for a long time, so we were delighted. We were told it was a top-secret job and they could not tell us more at this stage but would be told when we got to Perth.

On 6 January we flew in a Catalina to Rose Bay in Sydney and then by transport plane to Melbourne, Parafield, Ceduna, Forrest and Kalgoorlie, arriving in Perth the evening of 7 January.

Next day we had to report to Air Force House in St Georges Terrace, which was the headquarters for RAAF administration. We were duly advised that the job was to dispose of four Catalina aircraft, which had been flown by Qantas from Perth to Ceylon during the war.

Johnny Evans and myself went down to Nedlands where they were locked up and found they had been all been stored in the correct manner for out of the water. We got the motors going in the first one without much trouble. We checked hulls and everything to see if they were fit to fly and first ready was *Antares Star*. We all got down to Nedlands on 16 January and took off at 10.30am and did a two-hour test flight. Ted Withell was the skipper and Ted Hodson second pilot. We were told that an RAAF crash boat had gone out to Rottnest Island, and we were to liaise with them as to the water conditions, wind etc.

Scuttling No.1 *Antares Star*

So the next morning 17 January 1946 at Nedlands there was an RAF flying officer coming with us to certify the sinking of the Catalina and ensure nothing was taken out of the aircraft because he said that would break the Lend-Lease agreement. Also there was the key man for the whole exercise, a navy demolition expert by the name Peter Plowman. He came with us and we took off for Rottnest. After getting the weather conditions for the area from the RAAF crash boat and making our own observations, we decided it was OK for *Antares* to land. That was about 13 miles off Rottnest.

We landed, and the dinghy from the crash boat came alongside the Cat and we were all transferred to the crash boat. The night before, the explosives had been loaded aboard the crash boat. They consisted of two charges each of about 75 pounds of explosive. One was to be placed in the bow and one in the blister. Peter Plowman loaded the explosives aboard the dinghy and he rowed to the Catalina. He said there was no danger until these charges were connected. He unloaded them into the Catalina with the dinghy tied up at the blister. He then disappeared inside the Catalina, which was riding the swell and drifting with the wind. The plan was to detonate the first explosive in the blister, which would break the fuselage in halves then at a pre-determined time the explosive in the bow would go off and blow the front section into small pieces so nothing would be left floating. When he had set and connected the charges Peter climbed into the dinghy, grabbed the oars and rowed like hell back to the crash boat. He jumped on board. We tied the dinghy to the crash boat, went like blazes and stood off about half a mile from the Catalina. Peter was shaking a bit, and he admitted it was a hairy sort of a job.

The rear charge went off first, broke the Cat in halves and then the front charge went off. He said, "Good, it worked!". We stayed around to make sure it all sank so that there would be no shipping hazards. The crash boat, with us all aboard, set off back to Fremantle, up the Swan River to Nedlands. The first Catalina was duly sunk.

Johnny Evans stayed with me at my parents' house in South Perth. After a time, he moved to the YMCA in Murray Street, Perth, to see more of the city life. While we were in Perth, apart from our normal air force pay we were receiving a capital city allowance of 16 shillings per day so we were right up in the tourist class and as the three West Aussies had not been home for some time we did our best to stretch the job out as long as possible. After the sinking of *Antares* it was twelve days to 29 January 1946 before we test flew *Rigel Star*. The preparation of these grounded Cats was a long job and Johnny Evans and myself were the only two people who could say "yes" they were ready to fly.

Scuttling No.2 *Rigel Star*

So, on 29 January we took off and did a two-hour test flight in *Rigel Star*. Our take off was over the Canning Bridge and sometimes we would wonder whether we were going to clear it or not, but Ted Withell and Ted Hodson were very experienced pilots. Next day we did another test flight of two hours. Sometimes we would fly north along the coast and sometimes south, so all in all we had a good tourist view of our WA coastline. And then there was a break of four days until 4 February. I would mention that each day the RAAF crash boat would go out to Rottnest, stay overnight on board, we would go down to Nedlands each morning, get into radio contact with the boat and they would give us an indication of whether we should land or not. On 4 February we took off in *Rigel Star* and headed out to Rottnest. The sea conditions looked OK, so we set *Rigel* down and Peter Plowman and all the crew went through the same procedure again. This was 18 miles west of Rottnest. The rear charge went off as scheduled but the front charge did not go off. We thought what are we going to do now? The skipper of the crash boat said "No worries fellows, we'll fix it". So out came three 0.303 rifles with about 500 rounds of ammo and we stood off at a safe distance and started firing - it seemed almost impossible that we could put about 200 rounds into *Rigel Star* yet nothing happened! Finally, one shot set the front fuselage on fire. It blazed and blazed and finally all sank, so that was the exercise number two completed.

Scuttling No.3 *Vega Star*

It was not until 13 February 1946 that we took the next Catalina, *Vega Star*, out on a test flight of two hours. I may add that because these Catalinas had not been flown for a while the state of the hulls was an unknown quantity. But after a hull inspection following the Swan River landing of each test flight, we were fairly certain that they would stand up to a rough water landing in the open sea. Next day we took off at 1200 hours and flew *Vega Star* out 13 miles off Rottnest where the crash boat was waiting. We duly landed and went through the same procedure, and the exercise was completed.

Scuttling No.4 *Altair Star*

There was another ten days before we test flew *Altair Star* on the 24 February 1946. I do recall that we had a fair spell of bad weather about that time. Unfortunately, Ted Withell had developed malaria and could not fly so another pilot Flight Lieutenant Bill Swan replaced him. We did our two-hour test flight and then back to Nedlands and it was not until 27 February that the weather was right for the final mission.

On that day we flew out 13 miles west of Rottnest and sighted the crash boat and landed on the sea. Peter Plowman went through the same procedures once again. Set the two charges and the back one went off followed by the front one. But this time the outer section of wing with the wing float attached was blown off. The rest of the Catalina sank but this piece floated. It would have been a hazard to shipping, so had to be sunk. After a conference with all on board, it was decided that Arthur Jones and Johnny Evans would go out in the dinghy and sink it. The dinghy had the rope attached to the crash boat and we clambered and half fell into the dinghy. I was armed with a secret weapon and Johnny on the oars. I may add there was a fairly long swell of about four feet, and it was some time before all hands on the boat and dinghy got us close to the floating piece. You can imagine all the advice that was being thrown around as to what we should and should not do. As the floating piece was coming up, out came my secret weapon, the tomahawk from the crash boat toolbox. After a number of chops, I made one hell of a swipe and the axe got jammed in the float. Johnny yelled out "Let it go", and a voice behind us, the skipper of the crash boat yelled, "Don't let it go, I have to account for that". Anyway, the float was starting to sink, so I let the tomahawk go down with it.

I may add a point of interest. The crash boat used to stay overnight at Rottnest and local cray fishermen used to lay cray pots around the island and somehow or other on the first couple of trips we were all treated to a very nice meal of fresh crayfish cooked in the galley of the crash boat. Anyway, the fishermen had a good idea why their cray pots were empty and complained. The skipper got a blast from headquarters but, of course, it could not be proved. Anyway, we finished our job on 27 February 1946 and then we think that the powers that be forgot all about us until, suddenly, we got frantic telegrams to report to Air Force House in St George's Terrace. We were booked on a flight back to Sydney on 6 March. We took off from Perth at 1800 hours aboard an RAF Transport Command Liberator to Sydney, a nine-hour direct flight. The skipper's name was Flying Officer Horne and then on 8 March we got a flight from Rose Bay back to Rathmines on Catalina A24-10 piloted by Squadron Leader Grey.

In closing, I think if medals were handed out back then as they are today, Lieutenant Peter Plowman, Royal Australian Navy Demolitions, should have been awarded the highest valour medal available. It was a bloody hairy job and each time he did it I reckon he took his life in his hands.

Appendix 5

A QEA Lancastrian captain's log: Captain HB Hussey

Captain HB "Bert" Hussey had been flying with Qantas since 1934 and was highly experienced on DH.86s and Empires. He commanded eight RAAF Catalinas delivered by Qantas across the Pacific Ocean for the air force during 1941 (see Appendix 3). The following extracts from his logbooks give a good indication of Qantas pilot conversions on to the new Lancastrians and aircrew rostering for early commercial scheduled services on the new type. Of interest is his logging of Lancastrians by their British civil registrations in the early months, despite the aircraft being painted in RAF markings.

My thanks to Nigel Daw, President of South Australian Aviation Museum, for his assistance in accessing these logbooks, which were donated to the museum by Captain Hussey's family.

26 Apr 45	Lancastrian G-AGLS	RAAF East Sale	Training
28 Apr 45	Lancastrian G-AGLS	Sydney-East Sale	Training, night circuits
2 May 45	Lancastrian G-AGLS	East Sale-Sydney	Training
5 May 45	Lancastrian G-AGLT	Sydney-East Sale	Training
7-8 May 45	Lancastrian G-AGLT	East Sale-Perth-Learmonth	
12-13 May 45	Empire VH-ABG	Sydney-Townsville-Sydney	
22-23 May 45	Lancastrian G-AGLT	East Sale	Training, night circuits
29 May 45	Lancastrian G-AGLT	Ratmalana-Minneriya-Karachi	
2-5 Jun 45	Lancastrian G-AGLV	Minneriya-Learmonth-Sydney	
15 Jun 45	Lancastrian G-AGLV	Sydney	Pre-service test flight
16-17 Jun 45	Lancastrian G-AGLV	Parkes-Learmonth-Ratmalana	
24 Jun 45	Lancastrian G-AGLW	Ratmalana-Karachi	
30 Jun-1 Jul 45	Lancastrian G-AGLV	Karachi-Ratmalana-Learmonth-Sydney	
4 Jul 45	Lancastrian G-AGLV	Sydney	Test flight
6 Jul 45	Lancastrian G-AGLV	Sydney	Test flight
8-11 Jul 45	Lancastrian G-AGLS	Sydney-Gawler-Learmonth-Ratmalana-Karachi	
13-14 Jul 45	Lancastrian G-AGLU	Karachi-Ratmalana	
17-18 Jul 45	Lancastrian G-AGMB	Ratmalana-Minneriya-Learmonth-Sydney	
23 Jul 45	Lancastrian G-AGMB	Mascot-RAAF Laverton	2 hr 30 min night
25 Jul 45	Lancastrian G-AGMB	Laverton-Sydney	
7 Aug 45	Lancastrian G-AGLY	Sydney-Gawler-Learmonth	
11-12 Aug 45	Lancastrian G-AGLV	Learmonth-Ratmalana	
15 Aug 45	Lancastrian G-AGMG	Ratmalana-Karachi	
17 Aug 45	Lancastrian G-AGLU	Karachi	Test flight
18 Aug 45	Lancastrian G-AGLU	Karachi-Ratmalana	
21-22 Aug 45	Lancastrian G-AGLS	Ratmalana-Minneriya-Learmonth	
24 Aug 45	Lancastrian G-AGMB	Learmonth-Sydney	

28 Aug 45	Lancastrian G-AGME	Sydney	Pre-service test flight
31 Aug 45	Lancastrian G-AGMB	Sydney-Learmonth	
1 Sep 45	Lancastrian G-AGME	Learmonth-Ratmalana	
5 Sep 45	Lancastrian G-AGMC	Ratmalana-Karachi	
8 Sep 45	Lancastrian G-AGLT	Karachi-Ratmalana	
11 Sep 45	Lancastrian G-AGLX	Ratmalana	Test flight
12-13 Sep 45	Lancastrian G-AGLX	Ratmalana-Minneriya-Learmonth	
14 Sep 45	Lancastrian G-AGLU	Learmonth-Sydney	
28 Sep 45	Lancastrian G-AGME	Sydney	Pre-service test flight
2 Oct 45	Lancastrian G-AGME	Sydney-Gawler-Learmonth	
4 Oct 45	Lancastrian G-AGLY	Learmonth-Ratmalana	
7 Oct 45	Lancastrian G-AGMA	Ratmalana-Karachi	
11 Oct 45	Lancastrian G-AGLU	Karachi-Ratmalana	
13 Oct 45	Lancastrian G-AGMB	Ratmalana-Minneriya-Learmonth	
18 Oct 45	Lancastrian G-AGLZ	Learmonth-Sydney	
1 Nov 45	Lancastrian G-AGLW	Sydney	Pre-service test flight
3 Nov 45	Lancastrian G-AGLW	Sydney-Gawler-Learmonth	
6 Nov 45	Lancastrian G-AGLY	Learmonth-Ratmalana	
11 Oct 45	Lancastrian G-AGMD	Ratmalana-Karachi	
13 Oct 45	Lancastrian G-AGMA	Karachi-Ratmalana	
15 Nov 45	Lancastrian G-AGMC	Ratmalana-Minneriya-Learmonth	
18 Nov 45	Lancastrian G-AGME	Learmonth-Sydney	
28 Dec 45	Lancastrian G-AGLV	Sydney	Pre-service test flight
29 Dec 45	Lancastrian G-AGLV	Sydney-Learmonth	
2 Jan 46	Lancastrian G-AGLY	Learmonth-Negombo	
4 Jan 46	Lancastrian G-AGMD	Negombo-Karachi	
7 Jan 46	Lancastrian G-AGMC	Karachi-Negombo	
10 Jan 46	Lancastrian G-AGMH	Negombo-Learmonth	
13 Jan 46	Lancastrian G-AGLX	Learmonth-Sydney	
26 Jan 46	Lancastrian G-AGMB	Sydney-Gawler-Learmonth	
29 Jan 46	Lancastrian G-AGLT	Learmonth-Negombo	
1 Feb 46	Lancastrian G-AGLT	Negombo-Karachi	
4 Feb 46	Lancastrian G-AGLY	Karachi-Negombo	
4 Feb 46	Lancastrian G-AGMH	Negombo	Test flight
9-10 Feb 46	Lancastrian G-AGMH	Negombo-Cocos-Perth	
11 Feb 46	Lancastrian G-AGMF	Perth-Sydney	
24 Feb 46	Lancastrian G-AGLW	Sydney- Parkes	Training/check flights
25 Feb 46	Lancastrian G-AGLW	Parkes-Sydney	
26 Feb 46	Lancastrian G-AGMG	Sydney-Perth	
28 Feb 46	Lancastrian G-AGLW	Perth-Cocos-Negombo	
3 Mar 46	Lancastrian G-AGMD	Negombo-Karachi	
7 Mar 46	Lancastrian G-AGMF	Karachi-Negombo	
10 Mar 46	Lancastrian G-AGME	Negombo-Cocos-Perth	
13 Mar 46	Lancastrian G-AGMB	Perth-Sydney	

Appendix 6

A Qantas Lancastrian navigator's logbook

QEA Navigation Officer Arnold Easton's logbook provides a good indication of the route changes and flying times for the Lancastrian and Liberator operations during late 1945 and early 1946.

With thanks to Geoff Easton for permission to quote from his father's logbook.

Date	Aircraft		Hrs/min	Crew
21 Nov 45	Lancastrian G-AGMC	Sydney air test	1:23	Cpt Griffith, F/O Roberts, N/O Brearley R/O Turner, F/S Nock
22 Nov 45	Lancastrian G-AGMC	Sydney-Gawler-Learmonth	10:42	"
24 Nov 45	Lancastrian G-AGMB	Learmonth-Cocos-Ratmalana	13:40	"
28 Nov 45	Lancastrian G-AGME	Ratmalana-Bombay-Karachi	6:46	"
30 Nov 45	Lancastrian G-AGLX	Karachi-Ratmalana	6:50	"
5 Dec 45	Lancastrian G-AGMG	Ratmalana-Negombo-Learmonth	13:51	"
7 Dec 45	Lancastrian G-AGLZ	Learmonth-Sydney	9:50	"
14 Dec 45	Lancastrian G-AGMC	Sydney air test	0:53	Cpt Hoskins, F/O Kell, R/O Saxby. F/S Dannatt
18 Dec 45	Lancastrian G-AGMC	Sydney-Learmonth	10:22	"
20 Dec 45	Lancastrian G-AGMH	Learmonth-Negombo	14:15	"
23 Dec 45	Lancastrian G-AGME	Negombo-Bombay-Karachi	6:37	"
27 Dec 45	Lancastrian G-AGLW	Karachi-Negombo	7:02	"
29 Dec 45	Lancastrian G-AGLW	Negombo-Learmonth	14:17	"
2 Jan 46	Lancastrian G-AGMG	Learmonth-Sydney	9:16	"
15 Jan 46	Lancastrian G-AGMF	Sydney air test	1:06	Cpt Thomas, S/Cpt Morris, F/O Wharton, R/O McBean, F/S Kempson
17 Jan 46	Lancastrian G-AGMF	Sydney-Learmonth	11:35	"
19 Jan 46	Lancastrian G-AGMH	Learmonth- Negombo	14:57	"
23 Jan 46	Lancastrian G-AGLX	Negombo-Karachi	6:32	"
25 Jan 46	Lancastrian G-AGLW	Karachi-Negombo	7:05	"
29 Jan 46	Lancastrian G-AGMJ	Negombo-Cocos-Learmonth	14:06	"
1 Feb 46	Lancastrian G-AGLZ	Perth-Sydney	7:50	-
18 Feb 46	Lancastrian G-AGME	Sydney air test	1:06	Cpt Griffith, F/O Shannon, R/O R Young, F/S Wadick
19 Feb 46	Lancastrian G-AGME	Sydney-Perth	10:05	"
22 Feb 46	Lancastrian G-AGLX	Perth-Learmonth-Negombo	24:06	"

Date	Aircraft		Hrs/min	Crew
25 Feb 46	Lancastrian G-AGMA	Negombo-Karachi	7:00	"
28 Feb 46	Lancastrian G-AGMC	Karachi-Negombo	7:07	"
2 Mar 46	Lancastrian G-AGLZ	Negombo-Cocos-Perth	16:47	"
6 Mar 46	Lancastrian G-AGLY	Perth-Sydney	9:17	"
17 Mar 46	Lancastrian G-AGME	Sydney air test	1:13	Cpt Solly, F/O Morris, R/O Hatton, Sup R/O Robertson F/S Seaton & Thomas
19 Mar 46	Lancastrian G-AGME	Sydney-Perth	9:30	"
21 Mar 46	Lancastrian G-AGMB	Perth-Learmonth-Negombo	15:52	"
24 Mar 46	Lancastrian G-AGMA	Negombo-Karachi	6:51	Cpt Howard, F/O Morris, R/O Hatton, F/S Seaton,Thomas
28 Mar 46	Lancastrian G-AGMH	Karachi-Negombo	7:28	"
30 Mar 46	Lancastrian G-AGMJ	Negombo-Cocos-Perth	16:24	"
3 Apr 46	Lancastrian G-AGMC	Perth-Sydney	8:35	"
7 Apr 46	Liberator G-AGTI	Sydney-Darwin	10:01	Cpt Furze. F/O Fairservice, R/O Layton, F/S Seaton
13 Apr 46	Liberator G-AGKT	Darwin-Singapore	11:14	"
15 Apr 46	Liberator G-AGKT	Singapore-Darwin	10:24	"
17 Apr 46	Liberator G-AGTJ	Darwin-Sydney	11:34	"
3 May 46	Liberator G-AGTI	Sydney air test	0:53	Cpt Furze
3 May 46	Lancastrian G-AGMG	Sydney air test	0:57	Cpt Solly
5 May 46	Liberator G-AGTI	Sydney-Darwin	10.28	Cpt Furze, F/O Moreland, R/O Lanyon, F/S Pollard
10 May 46	Liberator G-AGTI	Darwin-Singapore	10:27	"
12 May 46	Liberator G-AGTI	Singapore-Darwin	9:14	"
15 May 46	Liberator G-AGTI	Darwin-Sydney	10:00	"

Sources

Aircraft Journey Logbooks, QEA Catalinas and Liberators, Qantas Library, Sydney.

Aircraft magazine editorial. *They don't rate medals*, Aircraft magazine, Melbourne, August 1946.

Airways. QEA House Magazine, various issues, Sydney, 1944-1948.

Ambrose, LR. *A Brief Outline of Indian Ocean Operations*. QEA Report, Sydney, 1945.

Ambrose, LR. *The Crossing*. QEA Report, Sydney, 1944.

Ambrose, LR. Interviewed by G Goodall, July 1977.

Baron, FC (flight steward). Diary of Qantas duty flights 1939-1946, Qantas Heritage Collection.

Bartsch, Howard "Joe". *Special and Urgent, Cats at War*, 2000.

Bennett-Bremner, E. *Front-line Airline*. Angus & Robertson, Sydney, 1944.

Brain, LJ. *General Report and Summary of Catalina Delivery Operations, 27 October 1941*, Mitchell Library, State Library of NSW, Sydney.

Brain, LJ. Letter to BOAC 19 July 1944, copy held by RAF Museum, Hendon.

Cowan, JLB. *Airline Pathfinders*, Airways, September 1948.

Cowan, JLB. Interview recorded by Greg Banfield, AHSA Aviation Heritage, Vol. 23 No. 3, Aviation Historical Society of Australia.

Crome, EA. *Qantas Aeriana*. Sutton Coldfield: Francis J Field Ltd, 1955.

Crowther, WH. *Operation of the Indian Ocean Regular Service by Qantas Empire Airways Ltd*. Sydney, QEA Report, September 1943.

Crowther, WH. Interviewed by G Goodall, July 1977.

Davies, REG. *History of the World's Airlines*. Oxford University Press, London, 1964.

Department of Civil Aviation files; QEA Catalinas and Liberators, National Archives of Australia

Director-General Civil Aviation (UK) telegram to Australian Government, 26 December 1944. Copy held by RAF Museum, Hendon.

East, Ron P. *A Flying Career*, self-published, Perth, 2004.

Fysh, Hudson. *Australia in Empire Air Transport*. Royal Aeronautical Society, London, November 1945.

Fysh, Hudson. Letter to selected QEA personnel, 30 April 1947.

Fysh, Hudson. *Qantas At War*. Angus & Robertson Ltd, Sydney, 1968.

Fysh, Hudson. *Australia in Empire Air Transport*. speech draft, November 1945.

Gillison, Doughs. *Royal Australian Air Force 1939-1942*. Canberra: Australian War Memorial, 1962.

Gray, RA. Aviation Heritage Vol. 33 No. 1. 2002, Aviation Historical Society of Australia.

Hughes, Allan. Aviation Heritage Vol. 29 No. 1 1998, Aviation Historical Society of Australia.

Labrum, Dick. Qantas radio operator recollections, unpublished account, courtesy Dick Labrum. A compilation was later published as *Long Hop Lancastrians* in Flypast magazine, February 1985.

Jackson, AJ. *British Civil Aircraft Since 1919*. Putnam, London, 1959.

Jackson, AJ. *Avro Aircraft Since 1908*. Putnam, London, 1965.

Jackson, Ken. Recorded interview by Greg Banfield, AHSA Journal, Vol. 21 No. 4, Aviation Historical Society of Australia.

Jackson, Paul. *East Indies Catalinas*. Aviation News, Volume 4, Number 1, (13-26 June 1975).

Labrum, Dick. *Qantas radio operator recollections*, unpublished account.

Livingstone, Bob. *Under The Southern Cross – The B-24 Liberator in the South Pacific*, Turner Publishing, 1998.

Lyndale, SM. *Catalina Conquest*. Wings 10 July 1944.

Mant, RO. Interview recorded by Greg Banfield, AHSA Journal Vol. 20 No. 4, 1980, Aviation Historical Society of Australia.

Miles, John. *Testing Time*, Neptune Press, 1979.

Minister for Defence papers, National Archives of Australia CRS A8161/1.

Morton, JG. *Flying for Qantas* 1945-1949, Aviation Heritage Vol. 30 No. 3, 1999. Aviation Historical Society of Australia

Moss, Peter W. *BOAC at War - Part 4*. Aeroplane Monthly, October 1975.

Neal, EH. *Order of the Double Sunrise*. Shell Aviation News, Nos. 413 and 414, 1972.

Neal, EH. letter to B Pattison, February 1978.

Peirce, Ivan. Interview, *Australians at War*, University of NSW, 2003.

QEA, *Non-stop Indian Ocean Service*, quoted in Airways, September 1944.

RAAF Historical Section. *Units of the RAAF: Volume 7 Maintenance Units*. 1995.

Roberts, Norm. Letter to B Pattison, July 1977.

Senior, Rex C. *The Double Sunrise Flight – My Story*. 2013.

Shelmerdine, R. No. 112 Air Sea Rescue Flight. RAAF report, Department of Defence, Air Historical Section: Canberra, undated.

Stroud, John. *Annals of British and Commonwealth Air Transport*. Putnam, London, 1965.

Taylor, PG. *The Sky Beyond*. Cassell, Sydney, 1963.

Tapp, RB. *Accident Report G-AGKU 16 October 1944*, Department of Civil Aviation Accident file, National Archives of Australia.

Tapp, RB. *Canton Island/Sydney Non-Stop*. QEA report, Sydney, undated.

Tapp, RB. *Landing at Cocos Island*. QEA report, Sydney, February 1944.

Tapp, RB. Letter to B Pattison, 1978.

Tebbutt, Geoffrey. *Kangaroo Service*, Airways, December 1944.

The Catalina Lament, copy held by Norm Roberts, 1977.

Winterbotham, FW. *The Ultra Secret*. Weidenfeld and Nicholson, London, 1974.

Yates, Albert E. Interviewed by Greg Banfield, Man & Aerial Machines journal, No.59.

Index of Personnel

Allan, Captain GU "Scotty" 17, 19, 35, 78

Allen, G 98

Ambrose, Captain Lewis R 10, 31-33, 41, 43, 45, 52, 57, 58, 67, 69-74, 81, 82, 97, 102, 103, 107, 108, 118

Anderson, RJ 118

Archbold, Dr Richard 14, 15

Bailey, JW 118

Baird, Arthur 28, 48

Balfour, Mr 117

Balfour, FJ 118

Balfour, HJ 118

Barker 59

Barker, Dudley E 10, 28, 29, 48

Barlow, Wing Commander 91

Baron, CF 78

Bartsch, Howard J "Joe" 42, 43, 53, 60, 61, 72, 118

Bennett-Bremner, E 20

Beresford, BPA 118

Bird, Captain Jack A 87, 118

Boyce, Captain FT 107

Brain, Captain Lester J 18-20, 23, 28, 53, 79, 82

Brearley, JH 118, 125

Brill, Wing Commander 68

Brown, Laurie 75

Bruce, SM 45

Buck, Captain RG 85

Bull, RD 118

Calwell, Captain KG 98

Capel, ND 98

Carr, RS 98

Carter, Gifford 109

Casey, RG 14

Chapman, FB 19

Christie, JF 98

Christie, R 98

Churchill, Winston 54

Clapp, FB 19

Clark, CW 118

Clarke, Warren R 60, 61, 72, 87, 118

Clunies-Ross, John 57

Cochrane, RG 49, 62, 118

Cole, Captain JH 108, 118

Colvill, Shinty 98

Compston, PH 118

Connolly, John 21

Cook, Captain 14

Corbett, AB 23

Corkran, TJ 118

Couchman, FA 118

Cowan, Jim LB 27, 39, 43, 58, 63, 83, 107, 118

Coward, Noel 54

Crabtree, Flying Officer Ed 68

Crowe, KM 118

Crowther, Meeta 104

Crowther, Captain William H 10, 12, 17, 21-23, 27-32, 34, 37-39, 41, 43, 46, 48, 49, 51, 54, 62, 63, 68, 71, 73, 82, 98, 103, 104, 118

Dannant, F/S 125

Day, John 49

Denny, Captain Orm 19, 66, 87, 88

Diegnan, Flight Lieutenant 91

Dobson, Jack 96

Duncan, Commander H 89

Dunne, RA 118

Dusting, Mr 87

East, Flight Lieutenant RP 93

Easton, Flight Lieutenant Arnold 89, 90, 125

Elliot, JG 118

Evans, Johnny 120-122

Evatt, Dr 78

Evatt, Mrs 78

Fairservice, RJ 72, 118, 126

Fielding, F/Lt 118

Foote, Charles LK 13

Fox, Wing Commander L 25

Furniss, Frank S 31, 37, 103, 118

Furze, Captain JAR "Allan" 55, 72, 95, 99, 118, 126

Fysh, Sir Hudson 23, 25, 26, 28, 29, 36, 39, 50, 51, 55, 62, 63, 67, 73, 81, 84, 102, 104, 109

Gairdner, Lieutenant General Charles 54
Gay, Ashley L 31, 60, 87, 118
Getker, Milton 74
Gillman, Mr 118
Grainger, D 118
Gray, Bob 106
Grey, Squadron Leader 122
Grey, Captain JL "Len" 66, 82, 118
Griffith, JF 118, 125
Grimes, Wing Commander Arthur 43, 103
Hale, KG 118
Hall, A 61
Harris, J 55
Hartley, J 118
Hatton, K 118, 126
Helfrich, Vice Admiral CEL 24, 25
Heymanson, Ernest 74
Hicks, PB 118
Hillgedorf, Mr 88
Hoare, George A 43, 118
Hocking, H 118
Hodson, Flight Lieutenant Ted 120, 121
Hogan, Squadron Leader John "Doc" 43, 60, 82, 103
Hollingsworth, F/Lt 118
Horne, Flying Officer 93, 122
Hoskins, Captain GP "Peter" 68, 118, 125
Howard, Captain SK 87, 89, 118, 126
Howse, Captain HT 28, 94, 118
Howson, PW 118
Hughes, Allan 100
Hurndell, CA 118
Hussey, Captain HB "Bert" 20, 21, 36, 87, 89, 118, 119, 123
Jackson, Captain Ken G 88, 89, 118
Jackson, R 90
Jackson, RJ 118
Jackson, RT 118
Jenkins, B 118
Jessamine, Colonel 60, 61
John, Flight Lieutenant 43
Johnson, Squadron Leader EL 103
Johnston, Captain Edgar C 30
Jones, Arthur W 65, 120, 122
Jones, Captain OP 67, 68, 70, 71, 73, 76
Keeling, Captain 57
Keith, JW 117, 118

Kell, F/O 125
Kempson, F/S 125
Kiernan, JT 118
Kingsford Smith, Sir Charles 13, 14
Knollys, Lord 67
Koch, AA 118
Koh, A 98
Labrum, Richard 90, 91
Lander, RE 118
Lanyon, GW 118, 126
Lawton, JH 90, 118
Layton, R/O 126
Lea, Squadron Leader 43
Learmonth, Wing Commander Charles C 76
Lee, HG 118
Leek, RN 118
Legg, David 34
Locke, HB 118
Lopes, AP 98
Louttit, LW 118
Lyon, Lieutenant Colonel 61
Lyons, D 118
MacArthur, Douglas 54
MacDonald, JD 117
MacKenzie, HS 118
MacKenzie, R 118
MacMaster, Captain Don F 41, 61, 62, 87, 118
Malcolm, DW 118
Mant, Captain RO 53, 89, 118
Martin, JR 118
Martz, Wally 61, 118
Maskiell, FJ "John" 41, 60, 118
Mather, MV 118
McBean, WD 96, 118, 125
McCaskill, FC 118
McClelland, NM 96, 118
McComb, AR 88
McCourt, TE 118
McLennan, CC 117, 118
Menzies, Prime Minister RG 16, 17
Miles, Squadron Leader John 88
Mills, Captain Harry G 60, 87, 118
Mitchell, FC 118
Moreland, G 118, 126
Morris, Captain ARH 107, 118, 125, 126

Morton, Captain John G 107, 108
Moxham, Captain JA 95, 118
Mueller, Captain 24
Muir, DC 118
Mullins, Captain 108
Mumford, Glen W 31, 37, 98, 103, 118
Murray, JN 118
Neal, EH "Ted" 39, 110, 117, 118
Nettleship, GM 118
Nicholl, Captain ER 78, 89, 118
Nicholl, Rex 88
Nixon, F 118
Nock, F/S 125
Nuske, AA "Dolph" 43, 96, 118
O'Dwyer, KA 118
Palmer, Captain E 89
Park, Air Marshal Sir Keith 92
Patterson, AS 18, 19
Patterson, CJ "Banjo" 43, 118
Pearn, Stan N 43, 118
Peirce, Ivan C 31, 49, 63, 118
Penny, Frank 88
Percival, Jack 15
Pirie, Group Captain Gordon 97
Plowman, Peter 120-122
Pollard, F/S 126
Pollock, Captain JC 68, 118
Porteous, R 96
Quirk, AJ 49, 118
Radford, HW 118
Ratford, Squadron Leader 90
Reeve, Captain FA 98
Rehfisch, Bill 68
Reijnierse, Lieutenant WJ 25
Richards, EC 118
Riordon, JR 118
Ritchie, Captain RJ "Bert" 31, 61, 62, 71, 78, 87, 89, 90, 103, 109, 118
Roberts, F/O 125
Roberts, Norm W 10, 20, 27-30, 39, 46-48, 56, 62, 71-73, 75, 118
Roberts, R 118
Robertson, S 118, 126
Rose, NS 118
Ross, HH 118
Rowe, J 118

Rumbold, Squadron Leader 31, 37
Sander, Frank E 43, 118
Saunders, Flight Lieutenant 91
Saxby, AM 118, 125
Scott, Wing Commander JC 27
Seaton, F/S 126
Senior, Rex C 30, 37, 40, 42, 49, 57, 58, 103, 118
Shannon, DS 118, 125
Shaw, Hudson 43
Shields, Captain PJR "John" 63, 94, 117, 118
Shirley, WA 118
Short, C 118
Sigley, Colin C 28, 46, 118
Simpson, A 118
Sims, Captain BC 99
Sims, Captain Eric 87
Skyes, WW 98
Solly, Captain John W 31, 51, 64, 71, 79, 118, 126
South, CW 37, 118
Standford, Admiral CS 89
Stewart, AW 118
Stewart, JR 118
Stitt, W 61
Stone, AK 98
Summerskill, Dame Edith 54, 55
Swan, Flight Lieutenant Bill 122
Swift, F/Lt 118
Tapp, Russell B 10, 21, 28, 31, 32, 36-40, 52, 58, 60, 62, 72, 73, 87, 97, 98, 103, 118
Taylor, Captain PG "Bill" 13-16, 18, 19, 21, 92
Thomas, F/S 126
Thomas, Captain OFY "Frank" 31, 33, 60, 61, 82, 87, 89, 96, 97, 101, 118, 125
Thunder, Wing Commander MD 27
Thurstun, GR 118
Tims, Major 61
Tracy, RDR 118
Turner, GW 118, 125
Ulm, Charles 13
Uren, RF 88, 118
van Prooyen, Commander 24
von Fritjtag Drable, Lieutenant 25
Wadick, F/S 125
Walker, BS 118
Walsh, ED 118
Wavell, General 23

Webster, Colonel 97
Weekes, L 72
Weston, RMH 118
Wharton, Alan 97, 118, 125
Wilhot, Flight Lieutenant 24
Willcox, WE 98, 118
Williams, Mr 87
Withell, Flight Lieutenant Ted 120, 121
Wood, J 118
Woods, H 118
Wright, DH 18, 19
Wright, GP 118
Yates, Captain Bert 87
Yates, Ron J 72, 104, 108
Young, R 118, 125

www.ingramcontent.com/pod-product-compliance
Lightning Source LLC
Chambersburg PA
CBHW060941170426
43195CB00026B/2997